Portland Comm

D0760729

The publisher gratefully acknowledges the generous contribution to this book provided by the General Endowment Fund of the Associates of the University of California Press.

A DIFFERENT NATURE

Kiki high in the trees at the Woodland Park Zoo, Seattle, 1980. (Author photo.)

A DIFFERENT NATURE

The Paradoxical World of Zoos and Their Uncertain Future

David Hancocks

University of California Press

Berkeley | Los Angeles | London

University of California Press
Berkeley and Los Angeles, California

University of California Press, Ltd.
London, England

Library of Congress Cataloging-in-Publication Data
Hancocks, David.
A different nature : the paradoxical world of zoos and their uncertain future / David Hancocks.
p. cm.
Includes bibliographical references (p.).
ISBN 978-0-520-23676-9 (pbk : alk. paper)
1. Zoos—History. I. Title.

QL76 .H35 2001
590'.7'3—dc21

00-053209

Manufactured in the United States of America
14 13 12 11
10 9 8 7 6 5
The paper used in this publication meets the minimum requirements of ANSI/NISO Z39.48-1992 (R 1997) (*Permanence of Paper*).

This is for Sam, and Tom, and Morgan.

It is dedicated to the memory of James W. Foster, D.V.M.

"These Mappin Terraces at the Zoological Gardens are a great improvement on the old-style of wild-beast cages," said Mrs. James Gurtleberry, putting down an illustrated paper; "they give one the illusion of seeing the animals in their natural surroundings. I wonder how much of the illusion is passed on to the animals?"

"That would depend on the animal," said her niece; "a jungle-fowl, for instance, would no doubt think its lawful jungle surroundings were faithfully reproduced if you gave it a sufficiency of wives, a goodly variety of seed food and ants' eggs, a commodious bank of loose earth to dust itself in, a convenient roosting tree, and a rival or two to make matters interesting. Of course, there ought to be jungle cats and birds of prey and other agencies of sudden death to add to the illusion of liberty. . . ."

H. H. MUNRO, 1919

The Toys of Peace. London: The Bodley Head.

CONTENTS

ILLUSTRATIONS

PREFACE

As a child growing up in the Welsh border country, in a world in which television existed but was unheard of by me, I could safely explore ancient oak woodlands, heathlands and marshes, and quiet narrow lanes, searching the internet of the more than seven hundred thousand miles of Britain's hedgerows, thick with wildflowers, that existed half a century ago. There were things to find in any season, familiar and rare: always rabbits, often a hedgehog snuffling for slugs, occasionally a bank vole or wood mouse, sometimes the soft blue eggs of a robin nestled in a cup of grasses in the deep shade of a hedge, their clarity of form and perfect smoothness contrasting with the tangled shelter of hawthorn and hazel. In springtime, there were newts with fire-red bellies and the magic of frog's spawn in clear shallow waters, lizards on dry-stone walls and grass snakes under the summer bracken, red admiral butterflies and yellow-banded snails, exquisitely handsome, feeding on bramble bushes on misty autumn afternoons. Once a year, on a trip to the seaside, I gathered seaweed, scrabbled for sideways-scuttling crabs, and dared to place a finger inside the sea anemones in the salty rock pools.

I had no knowledge of exotic creatures, not even the wolves and bears that once roamed Britain. Each day, however, I did gain an object lesson in ecology. My country-village home was an idyllic playground of fields and woodlands rich in wildlife. The town where I went to school, just a few miles northeast, lay in a region so blighted, so blanketed in industrial soot, it was called the Black Country. It had been the birthplace of the world's Industrial Revolution and was soured with chemical effluents and two centuries of grime and ash.

Once, the Black Country had been as green and as placid as any other part of rural England, but then its farms, hamlets, and villages were replaced with serried ranks of slum housing crammed into the gaps between iron mills, canals, collieries, blast furnaces, slag heaps, and railway lines. Wildlife had been extirpated, although rats, pigeons, thistles, sparrows, dandelions, and feral cats were abundant. Trees and children were stunted by lack of sunlight. Rivers and streams stank and steamed. The Black Country was, in microcosm, a harbinger of what we have been doing to the world ever since the Industrial Revolution quickened the pace of economic progress and introduced mass production and rapid exhaustion of natural resources, measuring success only in terms of financial profits. There was another omen, too. In the heart of the Black Country, a melange of exotic beasts lived in the Dudley Zoo: tigers, elephants, seals, baboons, monkeys, and bears, the flotsam of wild places from around the world trapped inside a high-walled enclave surrounded by a blasted and blackened land. I didn't go there as a child; I was a country kid. Zoos, like Dudley's factory rats and street pigeons, are an essentially urban phenomenon.

It wasn't until I was a university student studying architecture that I made a visit to London Zoo. There I saw exotic creatures for the first time: crocodiles, giraffes, mandrills, cassowaries, tropical fishes, and sea lions made a special impact on me. I cannot recall what I was expecting from the zoo, though I remember being curious and eager. I was not, however, anticipating the shock of seeing a gorilla. It wasn't his huge form that astonished me so much as the intelligence in his eyes. That, and the bitterly small size of his barren cage. This extraordinary animal, with his regal air, survived in a space no bigger than a garden shed. He was called Guy, and he sat on a concrete floor, soiled with his own excrement, looking out through bars and a glass window at a million people who shuffled past each year to gawk at him in his silent and solitary confinement. I walked away from London Zoo that day, as I have many others since, feeling confused and depressed.

It has been posited sometimes that if zoos did not exist they would have to be invented as places to save endangered species from extinction. Conversely, some people believe that no animal should be kept captive for any reason and that all zoos should be closed. Both of these extreme positions represent wishful thinking. Zoos are not the best places for holding and breeding rare species. Such an activity is better undertaken on large tracts of land where sufficient numbers of animals can be maintained for best genetic control, away from people, and in conditions conducive to their eventual release. As for abolishing zoos, the very strong roots of zoos as cultural attractions in our society make their forced closure an impossible goal. Sadly, even the worst examples of roadside atrocities attract paying customers. The implication in each extremist view, however, is correct: we should not accept zoos as they currently are.

My proposal is to *un*invent zoos as we know them and to create a new type of institution, one that praises wild things, that engenders respect for all animals, and that interprets a holistic view of Nature. It is possible to create captive situations in which wild animals can enjoy a life that is more comfortable, healthier, safer, and longer than they typically have in the wild. Moreover, though it is rarely achieved, we can present those animals in ways that reflect the splendor and wonderment of the wild. With a few changes we can design zoos that convey the richness of the natural world and that carry vital messages about our need to love, care for, and protect its diversity. Now almost totally separated from daily contact with Nature, people are quickly losing awareness of the importance of sharing the planet with a multiformity of living and essentially *wild* things. Zoos have the capacity to help us refocus our views of wild animals and wild places. They can encourage a new understanding of Nature.

But, sadly, when I lift images of zoos to mind, I find a jumble of unpleasant sights and sounds. Bored animals in small and sterile spaces, popcorn and ice-cream wrappers littering asphalt sidewalks, balloons, plastic snakes, panda keychains, hot dogs, artificially flavored drinks, chain-link fences, trees made of epoxy resin. I hear the echoes of clanging steel doors as lions and tigers and bears are locked away for the night and the reverberating screams of chimpanzees ricocheting off bare walls. I too easily find memories of small birds in impoverished cages, snakes coiled on gravel, living in a green-painted box with only a dish of water and a plastic vine, never able to stretch their body's length. Images come too readily of dusty enclosures, littered with steel feeding dishes, degraded with sawed-off tree stumps and rubber tires hanging on chains, bounded with endless lumps of fake rock walls. Plants seem to be relegated to the sidelines, except when gaudy splashes of color fill the flowerbeds in the ubiquitous municipal style of landscaping. The intensity of these semblances varies from place to place, but unintelligent design is commonplace. A casual review of the new zoo exhibits published each month in the design section of the American Zoo Association's newsletter, *Communiqué,* reveals a depressing eagerness to show off examples the crudeness of which should generate only embarrassment. All too often, zoos provide confused images of an artificial world, with their disjointed exhibits, second-rate food services, and wild animals held in ugly conditions unable to carry out the repertoire of their natural behaviors. Too many zoos are clumsy monuments to mediocrity. They enclose and confine the most exquisite masterpieces of evolutionary design in ugly and sometimes ludicrous environments, displaying and dishonoring beautiful creatures against backdrops of soiled brickwork and concrete.

But scattered among these memories are startling exceptions. The delight and astonishment of being close enough to hear the soft whiffling of a snow leopard, to watch the shuffling bulk of elephants rolling in a mud wallow, study a weaverbird busily interlacing grasses into his spherical nest, marvel at the crazy mating dance of cranes. I have seen groups of wild gazelles on the savannas of Zambia only at a distance, but at a few good zoos, I have found similar groups grazing on grassy plains and have gained extra pleasure from the knowledge that the zoo gazelles are unaffected by the scourges of parasites and will not suffer the agony of being chased down and eaten alive. On zoo visits I have seen people shed their irrational fears and discover by touching a snake for the first time that it is not slimy but smooth as silk and then come to recognize it as a wondrous example of biological engineering. I recall the delight in watching rehabilitated golden eagles soaring back into the skies after months of careful nursing by zookeepers and veterinarians. The surprisingly delicate slow-motion movements of hippos underwater, the deep whirring of a hummingbird's wings, the mesmerizing ballet of jellyfish, the flash of iridescent blue from the wings of a tropical butterfly that sat on my arm—these are personal experiences I would never have enjoyed without visits to a zoo. I have watched with pleasure as a docent convinced someone that tarantulas are worth welcoming rather than squashing. These glimpses of evidence that zoos can truly be places of wonder, bridges to paradise, sustain my often sinking opinion and soften my ambivalence.

With such a dichotomy of experiences, however, I find zoos a terrible challenge. They reveal the best and the worst in us and are stark portrayals of our confused relationship with the other animals with which we share this planet. It is illuminating, then, to examine our zoos, to untangle their muddled histories, and to ask whether they remain relevant. After thirty years in zoo design and management, I have formed some heretical opinions. Zoos are routinely justified by the four pillars of recreation, research, conservation, and education. Are these adequate?

The recreational aspect of zoos is surely suspect. Studies of zoo visitors have repeatedly shown that a substantial number attend simply as a family day out. Such indulgence is difficult to defend. Keeping wild animals in captivity warrants stronger justification than the setting for a social gathering.

Research provides a more contentious thesis. The proximity of zoo animals allows studies of their physiology and behaviors that would be dangerous and difficult in the wild. Behavioral research in zoos is the dominant activity, but is inherently problematic. The differences in their milieu make precarious any extrapolations with or comparisons between the behaviors of animals in wild habitats and those in zoos. And in any case, zoo staff rarely conduct research in the wild. There are, moreover, few trained scientists at most zoos, and data collection is usually for the purpose of solving captive-animal management problems, rather than contributing to the scientific literature. The matter of ethical research on zoo animals is also omnipresent. Invasive procedures are not warranted. It would seem that veterinary studies leading to improved physical care of the animals are the most valuable purpose for zoo research: which then raises the question of why the research is needed in the first place.

Conservation, in the form of breeding programs for zoo animals, is also a rather flimsy platform to support the continued existence of zoological parks. Fewer than five species have been saved from extinction by zoos, and some of them more by providence than prudence. Zoos are not, and for many reasons cannot be, sanctuaries for saving the world's wildlife: they deal with too few species and too little space for it.

This leaves education, which, in the original definition as a justification for zoos, probably meant a pedagogical approach. Monkeys have traditionally been represented in rows of cages, for example, so that people could make comparative observations of the physical form of different species. It is within a wider definition of education that the best and most viable reason for the continuing existence of zoos can be found. They have enormous potential to shape public opinion, to encourage sympathetic attitudes toward wildlife, and to educate the public about ecology, evolution, and wild animals. Zoos can open windows to a world of Nature that people could otherwise experience only via technology.

This potential role for zoos, however, is largely neglected. There is far more lip service than there are quantifiable results, more cant than can do. Of the millions of zoo visits that occur each year, few result in people exiting the zoo with better understanding of the inhabitants or more willing to make changes for the sake of wildlife. Indeed, some disturbing studies by Yale psychologist Stephen Kellert and Julie Dunlap of the Humane Society of the United States (1989) reveal that attitudes are more negative after a visit to some zoos. After people see animals in cages or in zoo exhibits that are highly artificial, they depart with "a significantly greater negativistic and dominionistic attitude to animals." That is why it is necessary to ask for more than improved aesthetics, better customer service, or healthier food and more sophisticated souvenirs. It is also essential to reach the point where the only zoos allowed by law are those that aim to create respect for wildlife and a desire to save wildlife habitat, by making animal welfare their first priority, by adopting conservation strategies as a central tenet of their operational, budgeting, and marketing decisions, and by injecting passion and daring into their interaction with visitors. I am confident that the public wants these challenges and wishes their zoos to have a strong voice and would welcome the leadership of zoos in wildlife and natural resources conservation. If these changes do not occur, then zoos must surely become increasingly meaningless.

There are stories strange and wonderful in the history of zoos, but perhaps their most extraordinary days are just unfolding. As an ever more ecologically hazardous future unfolds, our society needs institutions that can not only remind people of what losses are risked by reckless actions, but can also inspire compassion for other animals and reveal ways to live in better harmony with Nature.

The history of zoos is replete with contradictions. People have set up zoos because they wanted to control big strong animals and sought reflected power from being able to own savage beasts. But there are also zoo professionals who seek to inspire love and gentleness toward animals. And in recent years there are increasing numbers of people who want to work in zoos because they are passionate about wildlife conservation. Herein lies a critical aspect of the future for zoos. My ambivalence about zoos does not distract me from recognizing that we urgently need urban-based institutions that will carry not just the images but also fervent messages about the unnecessary and massive loss of wildlife habitats around the world, which is unsustainable and is an evil thing. We have no right and no need to destroy other life forms.

Of all the natural history–based institutions that we have invented—museums of geology, paleontology, zoology, and natural history; botanical gardens; arboretums; aquariums; and wild animal parks—it is zoos, I believe, that have the greatest capacity to adapt, absorb new functions, and amalgamate the content of other institutions. In this way, they can effectively carry the messages of conservation and wise stewardship. Zoos have the potential to present holistic philosophies with greater veracity and impact than any other type of natural history institution because they can present and interpret all parts of the story. Therefore, their historical focus on animals alone must shift and widen. The time is ripe for the zoo's metamorphosis.

More zoos are becoming habitat based, explaining ecosystems rather than only reciting facts about animals. Some zoos are beginning to develop exhibits that deal with concepts and ideas. Instead of seeing themselves simply as exhibitors of wild animals, a few zoos are learning to become storytellers. More will become involved in stories of deep history and of the interactions between human cultures and wild places. Sadly, there are zoos unworthy of the name that seek only profit and exploitation. Others are unfortunately limited to poor standards and have not shown themselves worthy of the magic and splendor of the animals in their care. They will not survive and do not deserve to. In this book, I examine the failures and the successes of zoos throughout their strangely checkered history, but more importantly I explore their amazing potential.

ACKNOWLEDGMENTS

I could not have written this book without the valuable help and kind assistance of many friends and colleagues.

The good counsel and encouragement of my literary agent, Lisa Ross, was an essential navigation aid, especially in the early days, but I would not have been able to find her in the first place without the kind assistance of Gary Luke; and it was David Brewster who was wise enough and who took time to put me in touch with him.

At various stages, Les Christidis of Museum Victoria, Mark Dimmitt and Gary Nabhan at the Arizona-Sonora Desert Museum, and Amanda Embury and Peter Temple-Smith at Melbourne Zoo read large sections of the manuscript. They each found errors and challenged me with questions or recommended changes that I found extremely useful. The few occasions when I chose not to incorporate their suggested modifications have possibly manifested themselves in errors or weaknesses; if so, I apologize to the reader and to them.

Specific information that I probably would not otherwise have discovered came from Jim Brighton, William Conway, Jo Gipps, John Gwynne, Devra Kleiman, Desmond Morris, Christopher Parsons, and Dana Payne, each of whom, despite very busy schedules, ungrudgingly gave me their time and attention.

Densey Clyne, Michael Dee, Lee Durrell, Beverley Flint, Jane Foster, Grant Jones, Barbara Logan, Carmen Lee, Terry Maple, Virginia McKenna, Christopher Parsons, Alex Rübel, Diane Shapiro, Ian Smith, Peter Struder, Ken Stockton, David Truex, Larry Vogelnest, and John Wedderburn greatly assisted me in obtaining illustrations. My attempts to locate sometimes obscure sources of information or to track down little-known references were generously aided by Kimberley Buck, Carol Cochran, Michael Dee, Warren Iliff, Stephen Johnson, Ian Jones, Kate Phillips, Michael Robinson, and most especially Val Hogan.

The technical skills and patience of Sandi Lehman, Jean Morgan, and Kim Woolley brought me through moments of panic and despair caused by Microsoft's word-processing system.

Practical assistance at critical times came from Barbara Dubow, Jackie Ogden, and Bonnie Kuykendall.

I am delighted to be able to offer my sincere gratitude and to formally thank on the printed page all of these people.

Several people at the University of California Press assisted with the production of this book. If their trades union was as powerful as those in the motion picture industry, they would all be named here, as they should be. Of those few whom I know, I would like to mention Danielle Jatlow, for copious assistance on numerous occasions. Ellen Browning, the copyeditor for the manuscript, applied polish in many places and brought many points forward for my consideration. Most of all I appreciated Suzanne Knott's grace and intelligence as she guided me gently through the bewildering maze of the publisher's world, and Doris Kretschmer's very careful attention to the details of final production.

CHAPTER ONE

COLLECTIONS AS STATUS

The elephant destined to become the most famous animal in the world was captured as a youngster, probably in Ethiopia in 1861, sold to a Bavarian animal dealer, sold again to the menagerie at the Jardin des Plantes in Paris, then exchanged for an Indian rhinoceros and shipped to London Zoo, where he arrived on 26 June 1865, half-starved, incredibly filthy, and covered with sores (Bartlett, 1900). His name was Jumbo, and his story reveals much about the dilemmas and peculiarities of zoos, especially of their modern history.

When Jumbo reached age seven, his keepers noticed a vast increase in his appetite, and he began to grow rapidly, on his way to becoming the biggest elephant ever seen in captivity. His daily diet included two hundred pounds of hay, two bushels of oats, one bushel of sweet biscuits, fifteen loaves of bread, three quarts of onions, occasional buckets of apples, oranges, figs, nuts, cakes, and candies, and enthusiastically taken hefty swigs of whisky (Preston, 1983). By 1880 he was more than eleven feet tall at the shoulder, and he soon became a great favorite of the British nation and earned enormous publicity for the zoo. His name became descriptive of anything unusually large and remains so.

In spite of Jumbo's star status, by the time he was twenty, the authorities at London Zoo were becoming increasingly concerned by what they de-

Jumbo at the London Zoo, about 1882. Matthew Scott is holding his trunk. The riders in the saddle include zoo superintendent Abraham Bartlett (wearing a top hat). His decision to sell Jumbo to Phineas Barnum created a national outrage. However, Jumbo had been displaying bouts of dangerous temper, probably, we now know, due to pain from impacted molars, and Bartlett was increasingly concerned about the possibility of a serious accident. (Photo © Zoological Society of London.)

scribed as his "fits of insanity" (Preston, 1983). They assumed that these dangerous bouts of violent behavior were the result of musth, a period when male elephants are in sexual rut. More than a hundred years passed before Richard Van Gelder, the curator of mammals at the American Museum of Natural History where Jumbo's remains are held, noticed while examining Jumbo's skeleton that the elephant had impacted molars. These teeth were likely erupting at the time that Jumbo began having his fits of violence that caused panic among the authorities at London Zoo. His unnatural diet, devoid of coarse grasses, leaves, and the roughage of bark, dirt, and roots, did not allow his molars to wear sufficiently quickly, and when his fifth pair erupted, they buckled and grew inward (Van Gelder, 1991). Almost certainly the excruciating pain from this caused Jumbo's temporarily aggressive behavior at London Zoo. Alarmed at the intensity of Jumbo's fits, zoo superintendent Abraham Bartlett sought approval for the purchase of a powerful rifle, as insurance. Of further concern to Bartlett was the fact that only one keeper at the zoo, Matthew Scott, was able to handle Jumbo, even though Bartlett had appointed him as Jumbo's keeper specifically "because he had no previous experience in the treatment and management of elephants" and would therefore, Bartlett hoped, more readily "attend to instructions" (Bartlett, 1898, in Vevers's anthology of the London Zoo, 1976). Scott, however, came to be intensely disliked by the zoo authorities. He was making extra cash by giving elephant rides and pocketing the money, and he apparently gloried in being uncivil to his social superiors. With the combined problems of Jumbo and Scott on his hands, Bartlett was delighted to receive and accept an offer in 1882 from the American circus showman Phineas T. Barnum to purchase Jumbo for ten thousand dollars (then about two thousand pounds sterling).

The British nation, however, was outraged, but with a fury fueled as much by jingoism as by compassion for the elephant. The potential impact of the relocation on Jumbo's welfare was mingled with concern about Britain's loss of a popular symbol of greatness. A public fund was set up to "Save Jumbo for the Nation," and the debate even reached Parliament. The editor of the *Daily Telegraph* sent a telegram to Barnum, requesting that he reconsider his purchase, explaining, "All British children distressed at elephant's departure." Barnum, the man who once declared, "Talk about me, good or ill, but for God's sake talk about me!" (Brightwell, 1952), saw Jumbo only as a money-making machine, and fanned the fires of publicity with vigor, replying, "Fifty millions of American citizens anxiously awaiting Jumbo's arrival." Publicity about Jumbo's departure was further heightened when several Fellows of the Zoological Society filed a lawsuit against the society's council for not calling an extraordinary general meeting to approve the sale of the elephant. The legal application, however, failed. Barnum's contract with the Zoological Society prevailed, and the day arrived for Jumbo's departure from the zoo. Now a new problem emerged: Jumbo refused to enter the shipping crate.

Day after day, futile attempts were made to entice, cajole, push, and drag him into the crate. There were rumors that Matthew Scott was giving the animal secret signs not to cooperate. The enormous shipping crate standing outside the elephant house for weeks became a new symbol for the nation; people began inscribing their names upon it as a mark of remembrance for Jumbo's supposedly patriotic defiance. Crowds left flowers; poems; gifts of dolls, other toys, and books; and food for the long journey. Attempts to walk Jumbo to Millwall docks, where a crane could hoist him aboard ship, had to be canceled when the elephant refused to pass through the zoo gates. Eventually Barnum came to realize that if he was going to get Jumbo he would also have to take Scott. He made a generous offer, which Scott immediately accepted, and a miracle seemed to happen that evening when Jumbo calmly walked into the crate.

The journey to the docks was a procession to match any royal funeral, some following the horse-drawn crate all the way from the zoo to make their mournful farewells. It is not surprising that Jumbo, his arrival having been preceded by so much attention, was a huge success in America. He arrived in New York on 9 April 1882, eleven feet six inches at the shoulder and weighing an estimated six and a half tons. Cheering crowds lined the streets on his journey from the docks to Madison Square. Jumbo mania hit America. Soon there were Jumbo cigars, fans, hats, jars of peanut butter, pies, and all manner of goods that benefited from being marketed as oversized. The *Philadelphia Evening Star* informed its fashion-conscious readers that the new shade of gray that spring was called "jumbo." No other animal's name has become so deeply embedded in our language.

Jumbo was paraded as a superstar in towns throughout the eastern seaboard; kept in the spotlight by the circus's insatiable publicity machine and traveling in his own Pullman Palace, Jumbo made a fortune for Barnum. On the night of 15 September 1885 in St. Thomas, Ontario, while being walked back to the circus train, Jumbo was hit by a runaway freight train. The train was derailed. Jumbo was dead. Barnum squeezed every bit of publicity from the tragedy. He sold the elephant's heart and bones to the highest bidders, respectively Cornell University and the American Museum of Natural History, and arranged for Jumbo's hide to be stuffed. Instructed by Barnum to show the animal "like a mountain," the young naturalist and taxidermist Carl Akeley stretched and overstuffed the elephant's skin to increase his already prodigious height by one foot.

For all his unique stature in zoo history and his extraordinary fame, in several important ways Jumbo typifies the story of too many magnificent animals in too many zoos over the centuries. Captured as a baby, his mother probably shot, sold to various institutions, never given proper care, put on display as a monstrous curiosity, reduced to a plaything to ride upon, moved from place to place, and haggled over in life and after death, Jumbo was both belittled and adored by the crowds that paid to see him. People wanted to see the massiveness of his form, yet saw too the superiority of their own kind. Awed and astonished by his size, they were also emboldened by the audacity of holding captive such a beast; humbled by his great bulk, they were yet prideful in the knowledge of human control over this giant. Jumbo's display brought out the best and the worst in people. Like many other zoo animals and like many zoos themselves, Jumbo satisfied the curiosity of all, the vanity of many, and the greed of a few.

INCONSISTENT ATTITUDES

The attitudes of humans toward wild animals seem always to have been hopelessly and perversely inconsistent. Around the world today, people adore, eat, fear, protect, worship, and, in laboratories, torture wild animals. The only constant is our inconstancy. The admiration that people have for wild animals is expressed by some in attempts to protect those animals and by others in attempts to shoot them for trophies. Some see the beauty of wild cats as a reason for keeping them alive in the wild, while it incites others to kill them for their fur. There are abundant societies for the protection of birds, but very few people seem concerned about cruelty to fish.

The history of zoos also contains perplexing and sometimes saddening events that reflect humankind's shifting relationships with wild animals, illuminating starkly varying perceptions in different times and places. Some people regard wild animals in captivity as subdued creatures with broken spirits, while others see the same animals as venerated beings that are protected and cared for. Indeed, zoo animals are often promoted as ambassadors for their wild cousins.

Zoos, from the most awful to the world's best, expose a perpetual dichotomy, which is the reverence that humans hold for Nature while simultaneously seeking to dominate it and smother its very wildness. They reveal both the best and the worst of human nature. The desire for close contact with wild animals is counterbalanced by the lure for ownership of things that intrigue. The wish to protect rare things is offset by a need for control.

For almost their entire history, zoos have been little more than gatherings of wild animals put on display to satisfy the dull gaze of the idly curious. If it were not so commonplace, it would be recognized that taking wild animals from distant lands and diverse environments, removing them from their natural social structures and habitats, and placing them on show in a city zoo, is a manifestation of arrogance rarely carried out for good cause. Even so, a common attitude of people who work in modern zoos is that wild animals are wonderful and require help and protection. Many of them passionately want to encourage a world in which human capacity for compassion and responsibility will lead to effective protection and wise stewardship of wild things and places. They see a choice between a world in which degradation of wilderness continues at alarming rates or one that keeps safe the wild homes of wild animals. The stark differences in these options is also, however, clearly evident in the inconsistencies of zoos, in the tenderness and the cruelty that has taken place in these peculiar institutions throughout history.

THE FIRST ZOOS

A fundamental shift in our relationships with wild animals (and one that has threads leading to the very idea of keeping wild animals in captivity) unfolded within the flickering light of the Paleolithic fire circle, when the first wolf ancestor of the dog scavenged for food scraps and started an association that led to its becoming the first domesticated animal. Humans quickly learned the benefits of owning animals. The dog, for example, provided many benefits: security, amusement, and play, greater hunting success and devoted companionship that invited the sharing of affection. Humans also learned something about themselves from this new partnership: those who had canine companions were distinguished from those who did not. Wild animal ownership bestowed prestige and power.

Ownership and display of possessions have long been symbols of progress in human society, and the first collections of wild animals by the socially elite represented significant power and distinction. Thus, for much the greater portion of their history, private zoos have served principally as important symbols of prestige, especially for the nobility, speaking status like living jewels. While few peasants could afford the indulgence even of a pet rabbit, a prince could cage the strength of the tiger, subdue the speed of the cheetah. Ownership of wild animals could signify a man's power over other men as much as his dominance over the beasts.

Zoos have evolved independently in all cultures across the globe. The first appeared about forty-three hundred years ago, in the Sumerian city of Ur. Wealthy Egyptian kings maintained collections that grew to thousands of wild animals, including monkeys, wild cats, antelopes, hyenas, gazelles, ibex, and oryx. These conglomerations were the source of much pride, for they were evidence of homage proffered by subordinate neighbors who sent gifts of wild beasts as marks of special obeisance. Around thirty-five hundred years ago, Tuthmosis III assembled a miscellany of exotic wild animals as a symbol of his imperial status. These living trophies paced the gardens of the temple at Karnak. His stepmother, Queen Hatshepsut, financed expeditions to gather wild animals for her royal collection. She sent a ship to Somalia, and it returned with monkeys, cheetahs, leopards, many types of birds, and even a giraffe to add grandeur to the gardens of her private palace. Rameses II had several giraffes in his zoo, as well as lions, one of which always accompanied him onto the battlefield. Rameses IX sent gifts of monkeys, crocodiles, and a hippopotamus to Tiglath-Pileser I, king of Assyria. A large pond was built for the hippopotamus at this royal zoo, where several species of large cats were kept in pits (Strouhal, 1992).

Whereas Greco-Roman and Judeo-Christian cultures would later come to perceive humans as superior to other life forms, ancient Egyptians held a view of closer kinship, seeing themselves as members of a whole family of plants and animals, in an eternal cosmic order. Some animals came to be regarded as incarnations of gods, from which royal families and whole tribes claimed descent. Baboons and monkeys were special favorites for keeping in temples. Deification of a species, however, brought dubious honor. Used in ritualistic sacrifices, sacred ibis, falcons, and crocodiles were mummified by the hundreds of thousands in sanctified cemeteries. The temple slaughters were so great that they led to extermination of these species in many parts of Egypt.

Wen-Wang, founder of the Chou dynasty some three thousand years ago, at the beginning of the classical age of China, built a zoo as part of a very large park. It was a peaceful, sacred place called, to the delight of modern zoo professionals, the Garden of Intelligence. As kingdoms became established across Asia, the libraries, museums, botanical gardens, and zoos in court palaces served as repositories of knowledge of the known world. Sometimes court favorites also had their private zoos. The courtesan Semuramis, in the Assyrian court of twenty-nine centuries ago, kept several leopards, and her son had a large collection of lions. Nebuchadnezzar, king of Babylon more than two thousand years ago, also specialized in keeping lions—the king of beasts, and the beast of kings.

While these early zoos were dramatic representations of their owners' social standing, the ancient Greeks revealed a different intent and a new perspective on animals. Plutarch made the first recorded declaration against neglect and ill-use of animals. Chastising the Roman statesman Marcus Cato for his petty avarice and lack of charity, Plutarch declared: "We should not use living creatures like old shoes or pots and pans and throw them away when they are worn out or broken with service." He lauded kindness and encouraged gentleness in dealings with other creatures. The Greeks collected assortments of animals and plants not just for show but for study and enlightenment. Twenty-four centuries ago, most Greek city-states maintained extensive zoos, and visits were an integral part of the education for young scholars. One of these students, Aristotle, later established his own private menagerie, the studies from which he compiled *The History of Animals,* the first zoological encyclopedia. He also became the first and for many centuries the only objective writer on natural history.

Alexander the Great had been one of Aristotle's students and maintained a lifelong interest in nature. He sent numerous plants and animals back to

Greece from his great eastern campaign, including a captive orangutan he found in India. It was in the city of Alexandria, founded in the year of Alexander's death, that one of his generals, whom he had appointed King Ptolemy I of Egypt, founded a zoo that became one of the wonders of its age. An indication of its size and complexity is revealed in a sumptuous procession to celebrate the Feast of Bacchus, when the zoo animals took the whole day to parade past the city stadium (Fisher, 1966). Preceded by musicians and dancers, one hundred and fifty men carried tree limbs "to which were attached wild animals of all sorts," while others carried gilded and painted cages of exotic birds, hawks, falcons, and giant snakes. There were ninety-six elephants drawing twenty-four decorated chariots, plus a dozen camels, a giraffe, and a rhinoceros. Sixty wild goats each also drew chariots bedecked with bright ribbons and bells. Seven pairs of wild asses in polished harness preceded three hundred exotic sheep, seven pairs of oryx, twenty-six white zebu, eight Ethiopian oxen, and eight pairs of ostriches, also in glittering harness. Then, heralded by festal horns and drums, came the big cats: twenty-four lions, fourteen leopards, and sixteen cheetahs as the grand finale to a magnificent parade (Loisel, 1912).

BLOODLETTING IN ROME

Jérôme Carcopino's study *Daily Life in Ancient Rome* shows how close contact with wild animals was a frequent occurrence for Roman citizens. Trained wild animals commonly performed in the city's bustling streets and crowded squares. Domestic and sometimes wild animals roamed many villas. Nero was just one of the many wealthy who liked to keep lions in their homes, as did Caracalla, who slept and ate with one of his favorites, Scimitar. He also took delight in introducing his pet lions at his dinner parties and found it especially humorous to release bears and lions into the bedchambers of his guests, snoring in their drunken stupors.

Most especially, however, Romans developed an unnatural appetite for lavishly funded bloodletting spectacles in the circuses. These shows included staged hunts, fights between animals, and fights between animals and people, including Christian martyrs, convicted criminals, prisoners of war, and professional gladiators. Toynbee (1973) notes that the first such show recorded in Rome was in 186 B.C., involving the slaughter of lions and leopards in a simulated hunt inside an amphitheater. Far from being disgusted, the populace clamored for more. Within less than two decades, the entertainment had increased to the point that one staged slaughter in Rome's Circus Maximus saw the death of sixty-three lions, forty bears, and an unrecorded number of elephants. Addicted to gore and sadism, the citizenry demanded ever more costly and elaborate spectacles of debauchery. The shows became so popular that people often camped overnight on the streets outside the amphitheaters to ensure the best seats.

The scale of the shows and the extent of the public killing was appalling. Titus had nine thousand wild beasts killed in A.D. 80. General Pompey financed a show featuring the slaughter of twenty elephants, six hundred lions, more than four hundred leopards, a rhinoceros, and some apes. Hadrian often had as many as a hundred lions killed in the Circus, and thirty-five hundred big animals were killed in twenty-six events staged by Augustus. Septimus Severus organized a show with an ingeniously designed ship that fell apart to expel into the arena a hundred animals, such as bears, lions, leopards, ostriches, bison, to be killed with spears, swords, and bows and arrows. This wretched event was repeated each day for a week. Trajan may have achieved the dubious honor of the greatest of these mass executions, when eleven thousand animals were killed to celebrate a military triumph. Commodus personally butchered a hundred bears, six hippos, three elephants, three rhinoceroses, a tiger, a giraffe, numerous ostriches, and uncounted lions and leopards. The record is dumbfounding.

The Emperor's Menagerie, just outside one of the main gates into the city, was partly a symbol of power, stocked as it was with exotic species from subject provinces and client kings near and far, even to the potentates of India, but also a way station for wild beasts en route to the killing grounds of the Roman amphitheaters. In a city that plumbed the depths of perversion to such an extent that Emperor Domitianus gave approval for a public play that ended each night with the torture and killing on stage of a condemned criminal (Carcopino, 1991), it is perhaps not remarkable that his brother, Emperor Titus, inaugurated the Colosseum with a gala of butchery that saw five thousand wild animals slaughtered in mock hunts or in fights to the death.

What does amaze is the stark discord between the intelligence required to build a monument so perfect in structure as the Colosseum, yet designed for such unconscionable cruelties. Toynbee (1973) notes that mosaics and painted friezes of the butchery were installed in the halls and living rooms of Roman homes. Clearly, intelligence and refinement are no guarantee for compassion or even civility (as Jennison records, 1937). Cicero seems to have been the only Roman to have ever spoken out against the madness, asking, "What pleasure can a cultivated man find in seeing a noble beast run through by a hunting spear?" It was the first and the last known public protest.

With much ingenuity the imperial architects created the impressive amphitheater of the Colosseum, large enough to house an audience of fifty thousand, for use as a public slaughterhouse. They engineered details for the spectacular shows with great technical skill. A ditch covered with metal grating, like a cattle grid, surrounded the arena, protecting the audience from the animals. The bears, elephants, rhinoceroses, tigers, water buffalo, lions, and other wild animals destined for death in these spectacles could be instantaneously launched from underground chambers into the football field–sized arena by an ingenious series of ramps, mechanical elevators, and hoists. The arena could even be flooded, to allow gladiators in boats to battle hippos, seals, and crocodiles. Giant fabric awnings could be attached to stone corbels to provide shade for spectators. The crowd often became frenzied from violent rushes of adrenaline, and as a precaution against stampedes, an elaborate system of staircases allowed evacuation in just minutes. Never was so much technical intelligence mixed in a building of such aesthetic harmony for purposes of such depravity and massacre. The spectator's eye could travel uninterruptedly from the rich blue of the Roman afternoon sky, down past the harmonious order of Doric, Ionic, and Corinthian columns adorning the three stories of arcades decorated with white marble statues, and come to rest on a scene of indescribable carnage in a pit of blood-saturated sand.

Emperors, politicians, generals, magistrates, and other wealthy individuals financed these horrific shows to gain personal notoriety. They were enormously expensive, with huge sums invested in capturing, shipping, feeding, and housing the vast numbers of animals. An efficient network of field men, distributors, and animal dealers supplied the Colosseum and other Roman circuses. The Caesars' slaughters were of such a scale and duration they eventually purged their domains of many wild beasts, eradicating the hippopotamus in Nubia, the lion in Mesopotamia, the tiger from modern-day Iran, and the elephant from North Africa.

In the countryside beyond the densely crowded and ever-noisy city of Rome, the world's first private zoos, established for vastly different purposes, became popular features of country villas. These personal zoos, perversely, were modeled on the Greek tradition of keeping a diversity of wild animals for the pleasure of learning by close observation. Meticulous details of these ancient zoos are recorded in Gustav Loisel's *Histoire des ménageries* (1912). Marine fish were kept in saltwater ponds. Large mammals were kept in palisades constructed of heavy oak timbers. Some of the private menageries were surprisingly large. Gordian III assembled a collection of thirty-two elephants, ten deer, ten tigers, sixty lions, thirty leopards, ten hyenas, six hippos, a rhinoceros, ten giraffes, and sixty wild horses.

Large aviary structures were often a specialty in these private zoos. Marcus Terentius Varro, at his country estate at Monte Cassino, built an aviary with several flight cages: one for songbirds, one for talking starlings, and others for peafowl. His dinner guests could delight in dining in a domed hall at the center of this complex, from where they could listen to and watch the birds on display in cages surrounding them. Similarly, Hortensus built a dining room overlooking an enclosure of deer and wild sheep and goats. Varro's aviaries, however, were more than a culinary diversion. He corresponded assiduously with other aviculturists, such as his wealthy friend Lucullus, the poet and epicure, who also built an indoor aviary that served as a huge dining room, and inquired eagerly about experiences with their collections, details of their breeding successes, and the status of their inventories (Fisher, 1966).

ZOOS GREAT AND GROTESQUE

Institutions such as libraries, botanical gardens, and private zoos collapsed and rotted with the fall of the Roman Empire. Very few menageries are found in the history of the long and barren eras of the Dark and Middle Ages. Istanbul had one of the few known zoos about fifteen hundred years ago that housed elephants, giraffes, and water buffalo in large enclosures, lions in cages, and monkeys in small pens. Nearly a hundred years later there was a zoo in Antioch, at the terminus of the city's main avenue, which was destroyed by the Persians in 538. Charlemagne, in many ways an exception to his times, kept three royal zoos in the eighth century. They housed monkeys, lions, bears, antelope, camels, many colorful birds, and, in 797, an elephant; most of them were gifts from Hārūn al-Rashīd, the Caliph of Baghdad, although one tame lion was sent courtesy of the pope. Charlemagne's animals were better fed and housed than most people of the day. The monastery at St. Gallen in Switzerland was another rarity, safeguarding an important collection of plants and animals that the monks diligently studied under exceptionally well-maintained conditions.

The thirteenth-century Holy Roman Emperor Frederick II, founder of the University of Naples and a diligent student of natural science, was the first person in several hundred years to assemble a major zoological collection in Europe, at his court in Palermo. It included elephants, a polar bear, a giraffe, leopards, hyenas, lions, cheetahs, camels, and monkeys. He wrote several authoritative books on birds, kept three permanent zoos in other Italian cities, took animals from his menagerie with him when he traveled, and often sent wild animals as gifts to other European rulers. His brother-in-law, England's Henry III, received some animals as a gift from Frederick, and in 1252 he set up a royal menagerie in the miserable confines of the Tower of London. The menagerie existed in that dreary place until 1834. Over the centuries, the Tower Zoo's fortunes fluctuated with the whims of successive monarchs. In the Tudor era, court guests were entertained there by staged fights between blinded bears and lions and tigers, cockfights, dog-fights, and contests between bulls and dogs. In its final days it continued this degenerate spirit: visitors in the nineteenth century could avoid payment of a cash fee by providing a small animal to be fed alive to the lions.

Emanuel I of Portugal was an unusually progressive ruler of the fifteenth century. He sponsored the expeditions of Vasco da Gama, Cabral, and Albuquerque. His court became a center of chivalry, science, and the arts, and he maintained a large and impressive aggregation of wild animals. Portugal was the first naval power of Europe and Lisbon was the gateway to Europe, which allowed Emanuel to control the market in exotic animals and to profit considerably from it. His zoo contained species unknown in other European cities, such as South American monkeys and macaws, plus an Indian rhinoceros, which was the subject of Dürer's famous etching and which died when the ship it was traveling in sank, en route to Italy, where it was to have been presented as a gift to the Vatican.

From the thirteenth through the fifteenth centuries almost every European prince and potentate had lions, maintaining the long-held symbolism of these animals as manifestations of power and strength. René, count of Anjou and king of Sicily, maintained a large menagerie in the fifteenth century at the Château d'Angers. Scattered around the grounds were a lion house, cages for small mammals, paddocks for ostriches and camels, ponds for waterfowl, cages for songbirds, and a large aviary. Early in the sixteenth century, Leo X, the Medici pope, kept tropical birds, lions, bears, monkeys, what was reported to be a snow leopard, and an elephant at the Vatican. This elephant, Hanno, had been given to the pontiff by Emanuel I, but caused panic and consternation upon presentation; obediently following instructions to kneel at Leo's feet, Hanno espied a stoup of holy water, dipped in her trunk, and sprayed the pope and everyone else in range.

The Fugger family, German bankers of immense wealth in the early sixteenth century, assembled great libraries and art collections and a considerable medley of wild animals for their private zoo in Augsburg, with specimens from Africa and the Americas. It was probably the grandest zoo in Europe at the time. Even so, it was dwarfed by the size and complexity of the most massive zoological collection of the era, and one of the largest ever, in Tenochtitlán (now Mexico City), where Hernando Cortés in 1519 discovered the staggeringly vast collections of Montezuma II.

Montezuma's extensive gardens were filled with an assortment of fragrant shrubs and flowers; one area was dedicated exclusively to medicinal plants. The adjacent zoological section included a phenomenally large and well-maintained collection of wild animals. Loisel records that a principal feature was a very large, airy house for jaguars and pumas, built of strong timbers and bronze bars, and decorated with animal sculptures. Next to it were enclosures for llamas, bison, and domestic species. Reptiles included caimans, turtles, iguanas, and snakes, the last restricted by water-filled ditches or housed in large boxes furnished with down nests for their eggs. Obtaining these feathers would not have been a problem, for birds seem to have been a specialty. There were ten fresh and saltwater ponds for waterfowl, aviaries for hummingbirds, parrots, and pheasants, and dozens of other bird species in a vast structure called the Bird Palace, with a separate house providing indoor and outdoor cages for birds of prey. That there was a team of nursing staff for sick animals and that keepers were required to go into the countryside each day to gather insects for additional nourishment to the birds' diets gives an indication of the quality of care for the animals in this zoo and the importance they held for Montezuma.

This zoological collection required a staff of six hundred keepers (male and female), and the birds of prey, cats, and snakes were so numerous they required a diet of five hundred turkeys a day. When he found this treasure, Cortés recognized its value for the Aztec empire and later made it a special victim of his campaign of violence. He laid siege to the city, and starving people ate many of the zoo animals. When Cortés finally took the city, he razed the zoo. Two decades before Europe created even its first tiny botanical garden (in Padua in 1543) and long before any European zoos approaching this stature were built, Cortés destroyed this place entirely, burning all the structures and slaughtering the remaining animals and their caretakers.

In the sixteenth century, Jelal-ed-din-Mohammed, the Mogul emperor of India known as Akbar the Great, established several zoos in various Indian cities, which also far surpassed in quality and size anything in Europe. Although he was illiterate, Akbar's court was a center for arts, letters, and learning, and he was a progressive legislator, eradicating slavery in his kingdom, abolishing the practice of forced suttee, pursuing liberal and eclectic religious policies, and forbidding animal baiting. His zoos reflected the tolerance and compassion of such endeavors. Unlike the cramped European menageries, Akbar's zoological parks provided spacious enclosures and cages, built in large reserves. Each had a resident doctor, and Akbar encouraged breeding and careful study of the animals. His collections, which included thousands of leopards, tigers, cheetahs, rhinos, deer, birds, a thousand camels, and more than five thousand elephants, were not assembled to assert his power and prestige but to instill love and respect for wildlife. Akbar's zoos were open to the public, and at the entrance to each he posted a message: "Meet your brothers. Take them to your hearts, and respect them."

It was to be a long time before such sentiments gained support in Europe.

Admittedly, by the late sixteenth century, royal zoos had come into great favor all over Europe, and virtually every king and prince owned at least one private menagerie. They were the source of much competition, and in the process some gained quite striking sophistication, with aviaries, seal pools, and large and elaborate animal houses. Their purpose, however, was rarely to encourage enlightenment but only to substantiate the power and rank of their owners. The Dutch were an exception and in the sixteenth century followed the Greek notion of building zoos as places of study open to the public for a fee. All others in Europe and for the next two hundred years were private menageries, owned by royalty or very wealthy individuals, maintained as expensive and exotic reflections of their social dominance.

In 1493 Columbus had returned with exotic species never before seen in the Old World: the first trickle in what was to be a flood of new plant and animal introductions from all over the globe. From the sixteenth to the eighteenth centuries more than one hundred *families* of animals were introduced into Europe. Plainly, there was confusion about the place of origin for many of them: macaws, Muscovy ducks, guinea pigs, and turkeys were named after Macao, Moscow, Guinea, and Turkey, respectively, but were in fact all indigenous to the Americas. Rich merchants, bankers, and princes of seventeenth-century Europe were able to add llamas, agoutis, South American parrots, hummingbirds, cotingas, honeyeaters, tanagers, birds of paradise, and chimpanzees to their flourishing collections. European menageries of the eighteenth century saw their first penguins, emus, rheas, condors, trumpeters, cockatoos, jacamars, toucans, cardinals, kangaroos, lemurs, marmosets, spider monkeys, anteaters, sloths, armadillos, raccoons, tapirs, and, but briefly, dodos.

A flush of new wild-animal collections spread across Europe as global exploration, economic expansion, and intellectual curiosity all intensified and gathered speed. With their new wealth, new knowledge, and abundant new animals and plants, Europeans also began to explore new ideas and to seek fresh perspectives about the wonders of the natural world.

CHAPTER TWO

THE EIGHTEENTH-CENTURY CONCEPT

Early European explorers returned home with more than rare objects and amazing stories. They also carried back an expanded view of the world, and people began to see the world and to understand their place in it differently. Each age drifts its way through time, preoccupied to varying degrees with aspects of science, aesthetics, ethics, or religion, all modified and fashioned by philosophy. In Europe, between the mid–sixteenth century and the mid–eighteenth century, fundamental philosophies shifted from a supernatural to an objective view of the world, from an unquestioning authority of the Church to faith in investigative science. These were changes that would gradually but steadily lead to a new understanding of Nature.

Our modern zoological parks, aquariums, botanical gardens, arboretums, and natural history museums are in large part the product of the new views that materialized at the end of this era. It is evident in their content, organization, and layout and particularly in the fact that we have such disparate natural history institutions, each devoted to different segments of the natural world, each allowing only compartmentalized views.

By examining the reasons and processes that guided these shifts in thinking and perception, it can be seen that today's zoological gardens, though tapped into historical roots as deep as civilization, are in truth grafted onto a Eurocentric and essentially English concept that is only some two hundred years old. Whether that concept, format, and view of wildlife is sufficient for today's needs must be questioned.

The ideologies about Nature that blossomed so profusely in England at the end of the eighteenth century first budded in fifteenth-century Florence with a new spirit of inquiry and inventiveness, a tolerance for new thoughts, and the stirrings of new understandings about Nature. Notably, a partnership formed between the arts and sciences. Painters, mathematicians, architects, and scholars sought ways to more objectively see and measure the world around them. They learned literally to perceive in a new way and developed new techniques in perspective that allowed three-dimensional space to be presented on a two-dimensional surface for the first time. Artists started making detailed and accurate observations, trying to depict the world more realistically and making use of their local zoological and botanical gardens for the purpose. Fra Angelico painted botanical specimens with scrupulous care in his *Annunciation.* Gozzoli's *The Procession of the Magi,* painted in 1459, includes meticulous representations of camels and cheetahs from the Medicis' menagerie in Florence. (Giotto did not have the benefit of a local zoo when he painted his groundbreaking masterpiece of realism, *The Adoration of the Magi,* 155 years earlier. He relied upon descriptive accounts of camels, which resulted in their having blue eyes, donkeylike ears, and horses' hooves.) Leonardo da Vinci made diligent studies of monkeys, bats, snakes, and lizards at the zoo in Milan and maintained a small medley of wild animals himself. Albrecht Dürer visited his local zoos often and made his famous lion engraving at the zoo in Ghent in 1521. His even more superb rendition of an Indian rhinoceros was based on a sketch and description ("it has the color of a speckled turtle") by Moravian printer Valentin Ferdinand of the first such rhino in Europe, at the impressively large zoo of Emanuel I of Portugal.

Stimulated by seeing strange new animals from distant lands, listening to stories about discoveries by explorers, and witnessing the opening of the heavens by astronomers, people in Western Europe began to change their views metaphorically as well as physically and to form new attitudes about the world around them. Also they began to examine anew their own place in Nature.

ATTITUDES OF RELIGION AND SCIENCE

The idea that everything exists for humans' sake has been common in many cultures, and Christians are not the only ones to have declared that their God gave humankind dominion over all other living things. The depth of prejudice we see in Christian history is nonetheless sometimes breathtaking. Medieval Europe was a closed, rigid, hierarchical society, convinced that everything in Creation had an ulterior aim: the edification and instruction of the sinful human (see, for example, T. H. White's *The Book of Beasts*). These attitudes persisted even into the age of Enlightenment. Keith Thomas's wonderfully readable study *Man and the Natural World* includes many examples. An English preacher in 1696 observed that God had most wisely located very dangerous animals in uninhabited regions "where they may do less harm." In 1705, physician George Cheyne declared that God had considerately arranged for horse excrement to smell sweet because He knew men would often be in its vicinity. A sixteenth-century bishop, James Pilkington, believed savage beasts were put on the earth as useful sparring partners for warriors. Whereas individual bears, wolves, lions, and other wild species looked alike, the Divine plan ensured that horses, cows, dogs, and other domestic animals came in convenient packages of varying colors and types so that people "may the more readily distinguish and claim their respective property." A Virginia gentleman, William Byrd, in 1728 guessed that horse flies had been created "so that men should exercise their wits and industry to guard themselves against them," and an English reverend decided that the louse's role was to encourage humans to keep clean. Clergyman Philip Doddridge noted that the instinct of schools of fish to swim into shallow waters "seems an intimation that they are intended for human use." One theologian noted that apes existed "for man's mirth"; singing birds were devised "to entertain and delight mankind." Cattle and sheep, it was decided, were given life so that their meat would be fresh "till we shall have need to eat them."

Not surprisingly, animals had no rights. (They were, however, subject to the same laws as humans and taken to court in many European countries. In 1457 in France, a pig was accused of killing five-year-old Jehan Martin, found guilty, and condemned to the scaffold.) When the vicar of Shiplake in Oxfordshire preached the first sermon against cruelty to animals, in 1772, he provoked universal outrage among his congregation who regarded it as a prostitution of the pulpit and proof of the vicar's growing insanity. A warmer welcome had been given to Bishop Ezekiel Hopkins's dictum on animals in 1692 that "We may put them to any kind of death [we] require." His contemporary, the Very Reverend Isaac Barrow, gained favor by declaring the common practice of vivisection to be easily excusable as "a most innocent cruelty," while clergyman Thomas Fuller preached in 1642 that "Christianity gives us a placard" to engage in bearbaiting and cockfighting. More recently, a nineteenth-century Jesuit, Joseph Rickaby, wrote, "We have no duties of charity, nor duties of any kind, to the lower animals, as neither to sticks and stones." Pope Pius IX refused his assent to an animal-welfare office in the Vatican on the same grounds.

By contrast, some other religions established philosophies deeply rooted in compassion for animals and protection of their habitats. The Bishnois, for example, are a Hindu sect living in India's most arid zone, on the edge of the Thar Desert. They allow neither the killing of wildlife nor the cutting of trees. As a result, wild animals, unafraid of humans, flourish in the desert scrublands around their villages. Yet when early travelers returned to England and told of the respect for life held by Jains, Buddhists, and Hindus, their reports were received with incredulity, even contempt. "Unaccountable folly!" thundered one seventeenth-century observer. "A discouraging impediment to the empire of man over the inferior creatures," noted another (Thomas, 1983).

The Bishnois faced a particularly severe test of their faith some 250 years ago when a maharaja sent troops to cut timber for a new palace. The villagers defied the soldiers by wrapping their arms around the trees and refusing to let go. Recently the women of Reni, in northern India, used a similar tactic and prevented foresters from denuding twelve thousand square kilometers of sensitive watershed. The Bishnois, however, did not have the protective eye of global television to save them and 363 men and women were slaughtered, axed down as they embraced the trees, before the maharaja learned of their resistance and ordered the troops to withdraw. In honor of their courage and commitment, he commanded that all his other subjects respect the Bishnois's principles, and he banned all hunting and all cutting of trees in the region. Today, a profusion of wildlife lives in this extremely arid zone, as well as healthy herds of cattle and goats, and a swath of sturdy thorn acacias and other shrubs and trees make the sand dunes green. The Bishnois, in addition, maintain a good standard of living. They are healthy, as is their environment, because they care passionately for the world around them.

Other religions, however, are not without their own myths about God-given authority for human domination of nature. The Maya, the Romans, and the ancient Chinese all proved quite capable of arrogant and large-scale destruction without the benefit of biblical instruction. Now that the Iron Curtain has disintegrated, we can see that state atheism is no guarantor of intelligent stewardship either. In the 1930s, Marxist philosophers considered wild animals and their habitats to be dispensable; they could be exterminated in favor of living space for people. Some argued that it was time to put an end to Nature (Cartmill, 1993). It is no wonder that Communists of the Soviet bloc have left a legacy of waste, ruin, and degradation of the natural environment beyond comprehension. The doublespeak of today's "Wise Use" movement is equally suspect and is in some ways reminiscent of attitudes prevalent in seventeenth-century Europe, when the reason to study Nature was simply to better know the enemy. A botanist in 1802 recommended careful study of insects "as we might thereby be enabled to find out the most certain method of destroying them." Horticulturists today would tend to approve. It remains a common view that Nature should stay beyond the garden wall, as any visit to a nursery will attest by the sight of shelves stocked full of pesticides, poisons, herbicides, and various other items of death and banishment. A gardening calendar for 1688 offered advice instantly recognizable today: "January, set traps to destroy vermin. February, destroy frogs and their spawn. March, the principal time of year for the destruction of moles. April, gather up and destroy worms and snails. May, kill ivy. June, kill wasps. . . ."

Botanical science originally had only utilitarian goals, motivated by the belief that everything on Earth existed to benefit humankind. The science of modern zoology had an equally practical genesis; in Britain, the Royal Society encouraged the study of animals to determine "whether they may be of any advantage." One aristocrat devoted himself to training pigs for labor. Many believed that tropical animals could be profitably introduced into Western agriculture. Europeans recognized that Nature had been lavish in producing many different animal species, but the distribution of this wealth had been capricious. Was it not logically part of man's task on Earth to correct this eccentricity? The original prospectus for the Zoological Society formed in London in 1825 was to acclimatize and breed new animals for domestication, and in 1860 a British Society for Acclimatization of Animals was formalized. La Société Zoologique d'Acclimatation, founded in Paris in 1854, devoted itself to the goal of "nothing less than to populate our fields, forests and rivers with new guests so as to increase our food resources." Similar societies were founded across nineteenth-century America. One had the intent to introduce to North America every animal mentioned by Shakespeare, and as a result starlings and sparrows are now prolific there. Acclimatization Societies in Australia were the origins of zoos in Melbourne, Sydney, Adelaide, and Perth. The first of these, founded in Melbourne in 1857, when that city was just twenty-two years old, was housed in the twelve-year-old Botanical Gardens. Oddly, this progressive combination of zoological and botanical gardens was not at all unusual in nineteenth-century colonial cities, but is extremely rare, yet more needed, today. These societies appeared in abundance in all of Europe's colonies, partly to introduce useful new animals but especially to establish familiar plants and animals from home. Thus homesick Britons introduced foxes, rabbits, blackberries, carp, starlings, and English songbirds to Australia, unwittingly unleashing enormous destruction of native species. Edward Wilson, one of the most ardent acclimatizers in colonial Australia, thought it important that "young lovers" on their evening rambles who "might occasionally fall a little short of topics for conversation" could benefit from the instructional value of "such material as the light of the glow worm and the song of the nightingale," both species that he was instrumental in introducing (Martin, 1995).

SUPERIOR BEINGS

Not just theological but also intellectual reasoning has placed humans above all other creatures. Francis Bacon, one of the brightest minds of the English Renaissance, was certain that "Man . . . may be regarded as the centre of the world, insomuch that if man were taken away from the world, the rest would seem to be astray, without aim or purpose" (Thomas, 1983). Aristotle had defined fundamental dissimilarities between people and animals, and Judeo-Christian teachings fused with his doctrines to represent humans not only as superior animals but as elevated beings midway between animals and angels. The differences were minutely categorized over the centuries, from Aristotle's notation that humans were political animals to Ben Franklin's description of humans as toolmaking animals. Their illogical climax was René Descartes's meticulous examination of animals as mere machines, like clocks. His explicit aim was to prove that humans were indeed the "lords and possessors of nature," and he set about the task with diabolical precision. Animals may be clever and complex, he acceded, but they are devoid of mind and soul and incapable of reason. Squandering his human ability to separate wrong from right, his moral sense suffocated by obtuse logic, Descartes came to the conclusion that animals were without sensation: automatons incapable of feeling. The screams he heard during experiments on living animals were to him the mere squeakings of machinery. Jean de La Fontaine observed one of these terrifying sessions: "They nailed poor animals up on boards by the four paws to vivisect them and see the circulation of the blood which was a great subject of conversation" (Krutch, 1961). One of Descartes's disciples, theologian and philosopher Nicolas Malebranche, questioned about this experiment, responded, "What about it? Do you not know that he does not feel?" (Spearman, 1966). So came intellectual support to the sadist: to the boot that kicked the dog, the whip that cut the horse, and the leg trap that cut to the bone.

The Cartesian view of man and nature occupied a central place in the minds of European intellectuals through the seventeenth and eighteenth centuries and generated a vast literature that continues to affect present-day philosophies. Descartes's theory that truth is to be found only by rational introspection, with mind and feeling kept resolutely separate, continues to bedevil present thinking. Al Gore is not the only one who believes that this philosophical model is at the root of our environmental degradation. Noting that "abstract thought is but one dimension of our awareness," he argues that "feelings represent . . . the link between our intellect and the physical world." Certainly it is difficult to imagine how people will maintain a healthy environment if they do not know and love the natural world. And we will not save endangered wildlife if we rely only upon an intellectual rationale. We need to love and respect creatures if we are to protect them.

Not all of Descartes's philosophies, however, found favor in England even in his own day. Henry More in 1648 told Descartes his was a "murderous doctrine." Most English philosophers derided Cartesianism as "contrary to common sense," and sought other avenues in pursuit of the fundamental variations between humans and other animals, for the English were just as eager to prove humanity's uniqueness and especially the immortality of the soul. One of the most universal and persistent yearnings remains, as Simon Schama expresses it, "the craving to find in Nature a consolation for our mortality."

While the philosophers' learned debates were restricted to relatively small audiences, their various topics became an integral part of society and lodged at the very foundation of public attitudes. Thus good manners, refined behavior, and civility came to be seen as indicators of human superiority. Dutch humanist Erasmus wrote a treatise on good manners, which stressed the differences between the way humans should act compared to animals: "Don't lick the dish, like a cat, don't neigh like a horse when you laugh, and don't bare your teeth, like a dog." Even today essentially human vices are unhesitatingly described as "brutish." Tabloid newspaper reporters routinely label sadistic criminals as "animals," ignoring the fact that no animal ever acted so cruelly as the human kind. British judges especially seem to enjoy using the term against louts and philanderers, expressing their frustrated belief that a bit of decorum would resolve these social problems. After all, as Henry Fielding explained in the eighteenth century, it was the dancing masters who taught "what principally distinguishes us from the brute creation." Or, as is carved over the entrance to Winchester College, England's first independent school, "Manners Maketh Man."

The similarities between people and animals have caused much anguish. New England clergyman Cotton Mather was urinating against a wall one day when he noticed a dog engaged in the same activity. "Thought I: 'How much do our natural necessities abase us, and place us . . . on the same level with the very dogs!'" Determined to define himself as "a more noble creature," he resolved that when engaged in similar future activities he would form "some thoughts of piety, wherein I may differ from the brutes."

PROBLEMS WITH CLASSIFYING

The firmness of the line that was perceived to separate humans and all other creatures justified such activities as hunting animals for sport, dissecting living animals for study, and exterminating any competing life forms. The definers of the disparities between man and animal also, by extension, cast a majority of the human race as bestial: children, with their untrained motor skills and undeveloped language; women, with their frightening menstruation, disgusting suckling of babies, and threatening sexual appetites; the squalid poor, the insane, the heathen Irish, and, of course, anyone of a different color. In short, everyone who deviated from the norm of the English gentleman. Nonetheless, the conceit that everything in the world was created for humanity was very gradually being eroded, weakened by the persistent investigations of science and philosophy. John Ray, who helped found in England what is now called the science of biology, joined the philosophers at Cambridge University to issue in 1691 the first challenge to the orthodox opinion that the world and everything in it, as Thomas Aquinas had insisted, was created only to serve humankind, arguing that all living things have natural rights (Krutch, 1961). Naturalists slowly began to develop more objective and less anthropocentric points of view and to ask challenging questions about human responsibilities to other animals and the human relationship to all of Nature.

Carl von Linné, the Swedish botanist, published his beautifully simple classification systems for plants and animals in the 1730s. Before the Linnean system, which designated two names—a genus and a species—to each different plant and animal, people had attempted to describe each form with lengthy and complicated Latin descriptions of what it looked like. Their difficulties were compounded by enormous confusion in attempting to comprehend the apparent chaos of Nature. What were the links? Where was the order? A tenth-century Chinese classification of animals by Tai Ping Kuang Chi included the following oddities (Graetz, 1995):

1. Those belonging to the Emperor
2. Tame animals
3. Suckling pigs
4. Stray dogs
5. The innumerable
6. Those drawn with a very fine camel-hair brush
7. Those that from a very long way off look like flies

Georges-Louis Leclerc, Comte de Buffon in the eighteenth century was no less mystified, declaring that the most natural classification of animals should be based on their usefulness to humans. In his system the dog and the horse received top billing. Other variations included classifications based on animals edible and inedible. Subdivisions could be based on aesthetics. Animals considered handsome or pretty (peacocks, owls) were categorized separately from those regarded as disgusting (lice, snakes, frogs). Birds were classified according to their singing abilities or their perceived demeanors: "generous" for the eagle, but "cowardly" for buzzards. Lions were "noble," ferrets "base."

A popular and long-standing assumption was that animals were ordered in some monarchical arrangement, with lions as king of the beasts, eagles the lord of the skies, and whales as monarchs of the seas, although Keith Thomas notes that scientists in late-seventeenth-century Europe debated at length whether the ape, elephant, beaver, or dolphin was at the apex of the animal hierarchy.

Linné's new system vaulted over all others, representing a new standard in how to learn and understand by closely observing. Instead of looking at the external form, Linné, starting with plants, looked instead to internal anatomy and based his system of classification according to differences in form and layout of the reproductive parts. Initially, it attracted prudish censure, because of its "licentious" emphasis on the sexual, and it was also vigorously attacked by some of the greatest scientists of the age: separating whales as mammals because of their reproductive system, for instance, instead of associating them with whale-looking fishes, seemed to flout common sense. It was deemed completely unacceptable in France, where the Jussieu brothers had already developed the idea of grouping plants based on natural affinities. Elsewhere in Europe the Linnean system eventually gained acceptance, and was generally accepted in England by the 1760s. Perfectly embodying the eighteenth century's desire for rational order and bringing clarity and simplicity to a perplexing problem, it also introduced a powerful psychological tool. The system helped change perceptions of Nature, greatly popularizing the study of natural history and especially botany. Now that people could conveniently classify all living things, in a logical method, this bred the confidence of control, of unlocking secrets. It came to signify the very essence of progressive science, the symbol of modern methods.

From all these changes emerged, most powerfully in England, a never-matched period in which people of all persuasions, from atheists to Christians, and of occupations that included scientists, artists, philosophers, poets, and essayists, obsessed themselves with developing an understanding of Nature and especially of humans' place in the natural world. It was a thesis with ancient roots, from Aristotle and Plato, that had long lain dormant. Now awakened, it began to flourish.

In addition, the concept of life as a Great Chain of Being, with a connectedness among all life forms, received wide acceptance in the eighteenth century. This idea had first formed in the Middle Ages, as an attempt to explain the universe, but now it was elaborated into a moral code. The self-serving assumption that everything was created for humans very slowly gave way to a realization that every link in the chain existed for its own sake, with equal claims to existence. This was not based on mere sentimentality but on logic, and though it was the outcome of philosophical speculation, not of scientific inquiry, its conclusions were not unlike those of modern ecologists when they write of disturbances to the balance of nature. Alexander Pope's eighteenth-century observation: "From Nature's chain whatever link you strike / Tenth, or ten thousandth, breaks the chain alike," is not essentially different from Rachel Carson's mid-twentieth-century warnings in her book *Silent Spring* about the dangers of severing ecological linkages.

Following this shift in perception, many writers decided, astonishingly, that humankind did not represent the apex of creation. If the cosmos was to be regarded as complete, it must, they deduced, contain other beings on other planets. Immanuel Kant came to the conclusion that beings on planets closer to the sun would be "far below the perfection of human nature," while those on more remote Jupiter and Saturn would be "lighter and finer" in every way. The degree of possibility for this, he said, "falls little short of absolute certainty" (Lovejoy, 1936).

Some found the notion of a mediocre position in the universe a gloomy prospect, but for many there was consolation in the recognition that humans, with their all-too-obvious shortcomings, must be far from the best that Nature could devise. It should be noted that in general a great sense of optimism prevailed over eighteenth-century Europe and that it was based on acceptance of the way things were. If humans were not the most ideal form of life in the universe, and if the "nature of things" prevented us, and our world, from being perfect, then a reasoned acquiescence of the inevitable was called for. Seeing that life could not be otherwise made it all the more bearable. Nature, declared Tennyson, was red in tooth and claw. It could be neither worse, nor better.

These beliefs required intellectual acrobatics to equate the obvious imperfections of the universe with an all-knowing and infallible creator. William King, archbishop of Dublin, explained for example that whereas God *could* have created a world free of predacious carnivores, this would have resulted in a world less full of life. He reasoned that it was better for some animals to be given life "for a time, though they be devoured afterwards," and went so far as to declare that an animal being eaten by a lion "should rejoice, as does God, that in its demise it is allowing the lion to act according to its God-given nature." Gottfried Leibnitz wrestled with an equally contorted paradox: if imperfections were inherent in Nature, why did God not simply refrain from creating anything at all? "The abundance of God's goodness is the reason. He preferred that the imperfect should exist, rather than nothing. To have a thousand well-bound copies of Virgil in your library, to sing only arias from the Operas, to have all your buttons made of diamonds, to eat only partridges, to drink only of the wines of Hungary—could anyone call this reasonable?" (Fleming, 1974).

ENVIRONMENTAL EROSION

Another general unease accompanied these doubts and hesitations about people's place in Nature and their relationships with other animals. It was a discomfort that would later be salved by the enormous public pleasure in using urban botanical and zoological gardens as a retreat and for perceiving them as microcosms of Nature in the cities of the nineteenth century. The progress of civilization in the eighteenth century was rapidly creating dangerous, dirty, and ugly urban areas. At the time of the Renaissance, England had a population of three million people and about eight million sheep: a green and quiet agricultural country with four million acres of deep forests. London had a population of about seventy thousand. Towns were small, tidy, and contained. Exeter, for example, contained only ninety-three acres within its walls. No one lived more than a fifteen-minute walk from open country. In this environment, urban life was synonymous with civility, manners, learning, security, and intelligent design. Two centuries of growth and crowding changed that view. Towns and cities rapidly became polluted, full of stench, clamor, and disease. In 1743 a writer for the *London Magazine* complained of living in an environment "immers'd in smoke, stunn'd with perpetual noise."

Relentless progress had corrupted the townscapes, and it had brought unwelcome changes to the countryside that resulted in a shift in the English view of Nature. Tamed and cultivated land had traditionally been the measure of satisfaction, equating fertility with beauty. The stiffly geometric gardens of Tudor and Stuart England were verification of human order upon the chaos of wild landscapes. Symmetry, be it of fruit trees or furrows, has long been a mark of civilization in the countryside, and until recent times, Europeans viewed the unfettered character of wild places with abhorrence. William Camden, a sixteenth-century historian, thought the Welsh landscape was hideous "by reason of the turning and crooked by-ways and craggy mountains." A contemporary description of the heathland of Surrey, "within seventeen miles of the capital city" (and now the most coveted countryside in England), rejected it as "horrid and frightful to look on, not only good for little, but good for nothing." Samuel Johnson was "repelled" by the "wide extent of hopeless sterility" in the Scottish Highlands, and James Howell, in 1621, thought the Alps "high and hideous" (Thomas, 1983).

Yet by the end of the eighteenth century, this perspective on Nature had changed and changed completely. Wild lands had become a source of spiritual renewal for the English, the wilder the more powerful their inspiration.

Within a century, mountains had changed from "monstrous excrescences" to objects of admiration. Anthony Ashley Cooper, the socially influential third earl of Shaftesbury, wrote in 1711 "how the Wildness pleases" and declared that even deserts had their "peculiar beauties." What caused such a turnaround? The reason was that the English countryside had suffered dramatic and widespread visual changes, caused by a rapacious expansion of agriculture. As more and more of the land was put to the plow and carved into the regularity of formal enclosures, the results sowed an opposing set of aesthetics. Nature was disappearing, and thus becoming more precious.

Increasingly, naturalness came to be the preferred guide for shaping the English landscape. Fences and walls were to follow natural contours rather than the strictures of geometry. Subtle placement was appreciated more than the convenience of straight lines, and a rural landscape was consciously created to satisfy both profit and visual delight. At the same time, and from the same concerns, a distinctly English style of gardening evolved. William Kent, Lancelot "Capability" Brown, and Humphrey Repton, the founders of landscape gardening, created irregular and romantic landscapes on a magnificent scale that were a conscious rejection of the excessively formal style that had been imported from France and Holland. Their work, described by one of America's best-known landscape architects as "one of the greatest artistic achievements of the West," formed a remarkably perfect partnership with the cultivated softness of the new English landscape (Olin, 2000).

The new English-garden style was accompanied by a fresh way of looking at plants, figuratively as well as literally, as some botanists began showing an interest in plants for their own sake. Local flora societies and botany clubs emerged, and by the early 1800s amateur naturalists had botanically codified most of England. The English began to find elegance in thistles, blackberry, furze, broom, ragwort, yarrow, rushes, briar, sloe thorn, and ferns. Common wildflowers found great favor with the Romanticists. Ruskin poured scorn on cultivars: "unfortunate beings, pampered and bloated above their natural size, [with] speckled and inharmonious colors . . . glaring away their term of tormented life among the mixed and incongruous essences of each other." Tennyson and Wordsworth pronounced similarly strong dislikes for hothouse plants. "Long live the weeds," wrote Gerard Manley Hopkins.

Suddenly, wild animals also had their champions. Serenely affluent England came to see them as potent symbols for the spiritual purity of wild places. This comfortable indulgence could be cultivated in Britain because no large dangerous animals had existed there for many hundreds of years. Wolves and bears had long disappeared. But it was also becoming apparent that many other wild animals were being lost. Acts of Parliament in the sixteenth century had started a rigorous campaign of extermination against any animal that posed even the slightest competition for resources. Crows, rooks, foxes, polecats, weasels, stoats, otters, hedgehogs, rats, mice, moles, hawks, buzzards, ospreys, jays, ravens, and even kingfishers all found their way onto official death lists. To exacerbate this colossal destruction, rapid land clearance for pasture over the preceding two centuries had eradicated millions of acres of wildlife habitat.

Some animals in England had always been protected as curiosities, like the wild Chillingham cattle introduced by the Romans, or for prestige, like the wild swans maintained at Abbotsbury since at least 1393, each technically owned by the monarch. For selfish reasons, royal hunting preserves had also long protected the wild habitats of deer, otters, hares, pheasants, and salmon, setting a regal example for groups such as today's Ducks Unlimited and Trout Unlimited.

In the England of 1800, however, when it was becoming clear that many wild species were disappearing, the first stirrings arose of a belief that Nature was worthy of respect and care for its own intrinsic value and that wild animals should be protected simply for their own sake. William Blake decided, "Every thing that lives is Holy." Naturalness had emerged as the eighteenth century's source of aesthetic delight in England's gardens and countryside, but it was untamed Nature that became the object of veneration. The reverence, moreover, was commonplace, as David Cannadine (1989) notes, "from the Chartists to the Fabians to the High Tory paternalists." With the very visible destruction of wild places throughout the island, and the consequent decimation of wildlife, a new set of ethics was sown. Nature came to be revered, praised by artists, poets, and philosophers. Rousseau was the greatest prophet of this passionate new devotion, declaring that the best way to cure the ills of a world overly refined and choked on progress was to turn to Nature and live as simply as possible. The principle of this idea would find favor again in the nineteenth century when America saw its wild frontier as the last bastion of individualism and came to believe that in its wilderness was the antidote to the poisons of a polluted industrial society. Initially, however, adoration of Nature became, more than anywhere else, an English phenomenon. The British, not the French or Spanish, botanized the Pyrenees. The British, not the Swiss or Italians, formed the Alpine Club. It is noteworthy that transplanted Britons—photographer Eadweard Muybridge, painter Thomas Moran, and naturalist and visionary John Muir—were among the most prominent promoters behind the establishment of Yosemite Valley as a national park in 1864. Notably, too, John James Audubon's *Birds of America* (1826–38) was first published not in New York, Boston, or Philadelphia, but in London and Edinburgh, and of the 180 subscribers listed in 1831, 152 were British. And it was in England that the idea was born of preserving uncultivated nature simply as a symbol of freedom and as an essential spiritual resource. Wordsworth was the first to note, "Wilderness is rich with liberty."

It was this English enthusiasm for wild landscapes and for informal gardening styles, growing from a wide range of social, intellectual, and economic causes, that had such significant effect upon the zoological park development that mushroomed first across Europe and then around the world throughout the nineteenth century.

CHAPTER THREE

THE NINETEENTH-CENTURY PHENOMENON

Despite the intense interest in England about places and things wild, the seventeenth and eighteenth centuries had seen no significant new zoos there. New zoological institutions were founded in many other countries during this period throughout Western Europe as well as Morocco, Egypt, Iran, India, and Russia. The Swedish Royal Zoo served as a living laboratory for Linnaeus's work in the scientific classification of animals in his *Systems of Nature,* published in 1758. Sadly, not all royal European menageries enjoyed such intelligent patronage, as James Fisher recounts in his *Zoos of the World.* Augustus II of Poland held contests at Dresden modeled on the Roman circus spectaculars, with fights between tigers, lions, bulls, bears, and wild pigs. One day in 1719, he allegedly personally shot every surviving animal in his collection. France's Henry III attended Mass one morning then afterward shot and killed every animal in his menagerie of mainly lions, bears, camels, monkeys, and parrots. He said he had dreamed that they were going to eat him.

Conversely, French king Charles X's affection for the animals in his Jardin du Roi was more representative of the personal interest many regal owners took in their living collections. When Charles received a giraffe as a gift from Muḥammad ʿAli Pasha, the Ottoman viceroy of Egypt, in the summer of 1827, he arranged for her to wear a cape embroidered with the French fleur-de-lis and the Egyptian crescent on her walk from the docks in Marseilles to Paris. This extraordinary journey, supervised by Étienne Geoffroy Saint-Hilaire, who had been director of the menagerie at the Jardin des Plantes, was the event of the year (Allin, 1998). It inspired songs, poems, and even towering hairstyles, so high that women had to sit on the floors of their carriages. The fashion color of the year was "belly of the giraffe," and men wore "giraffic" hats. Giraffe shapes and skin patterns appeared in and on parasols, brooches, tobacco tins, wallpaper, faience and Limoges plates, jewelry, soap, cake molds, and furniture. The influenza epidemic in Paris the following winter was known as "Giraffe flu."

The giraffe's winter quarters were quite elegant, with parquet flooring and the walls insulated with an "elegant mosaic" of straw matting: "truly the boudoir of a little lady," wrote Geoffroy Saint-Hilaire. Some of the new European zoos of the period were also quite grandiose and carefully designed, but none of them yet reflected the new English zoo style that was to blaze the pattern for a fantastic spread of zoo developments worldwide throughout the nineteenth century. Until then, the French pattern—formal, orderly, and symmetrical—was predominant. The most stylish zoological park of the era was created by Louis XIII at Versailles in 1624. It neatly reflected many aspects of its era, society, and owner-designer. A perfect example of the desire to master and exclude wilderness, its radial symmetry manifested the French academic approach to planning: the orderly display of animals showing the superiority of civilized humanity and the formal geometry of its plan revealing the finesse of its creator. An elegant structure at the center of the site, with elaborate sculptural decorations, walls delicately painted with birds, butterflies, and other animals, and a main entrance facing an *allée* leading from a classically proportioned lodge, perfectly symbolized the stately imposition of the king and his wishes upon the landscape.

Conspicuously designed as a potent symbol of regal opulence, the Versailles menagerie was an obvious target for the wrath of the proletariat and most of the animals were butchered when the menagerie was sacked in the very early days of the French Revolution. The National Revolutionary Convention approved construction of a public menagerie in 1793 on the site of the Jardin du Roi, renamed Jardin des Plantes (the original Jardin des Plantes had been established in 1635, as the Royal Medical Garden), and thereby created the world's first national zoo and the first open to the general public. Its founding occupants were the six remaining animals salvaged from what had been the very large assemblage at Versailles: a quagga (a now-extinct type of zebra—the last one on Earth died 12 August 1883 at Amsterdam Zoo), a hartebeest, a crested pigeon, an Indian rhinoceros, and a lion and dog that were inseparable companions. Later that year, a polar bear, a panther, and other animals seized by police from a traveling circus were added to the zoo's collection. It operated as a division of the National Museum of Natural History, which took a new approach to menageries. Two new chairs of zoology were established. Jean-Baptiste Lamarck, the biologist who first broke with the idea of immutable species and who promulgated the concept of evolution, was given the splendidly titled Chair of Insects and Worms. Bernard de Lacépède was to have received the equally loftily named Chair of Quadrupeds, Cetaceans, Birds, Reptiles, and Fish, but being of noble birth had to flee Paris. This was a critical loss. Lacépède had envisioned a departure from elegant iron enclosures at the zoo and advocated hidden barriers within naturalistic settings. The post was awarded to the twenty-one-year-old Étienne Geoffroy Saint-Hilaire in Lacépède's stead.

The menagerie did not thrive, however. Whereas the public clamored for a great variety of animals, Geoffroy Saint-Hilaire took less and less interest in the institution, devoting himself to traveling, writing, and philosophical studies. The place soon showed signs of neglect, and it continued to decline until Étienne's son, Isidore, succeeded him as director in 1838, bringing with him the conviction that he had the blueprint for the perfect zoo. He had found the inspiration for this new concept in Francis Bacon's book *New Atlantis.* This ideal zoo would encompass several scientific departments, aiming to create a zoo that would both educate and amuse the visiting public and would improve French agriculture by acclimatizing exotic species. But, as his father's had, Isidore's revolutionary interest waned, and he eventually abandoned his project.

Despite all this official activity, the French Revolution sadly led to no significant improvements or permanent changes for zoos. Instead, it was the French monarch's menagerie style with a dominant, central architectural feature as the hub for a wheel of radiating paddocks that was to influence all of the important European zoo developments for two centuries, just as the rotunda, the oldest building in the Jardin des Plantes, designed to replicate the cross of the Napoleonic Legion of Honor, made its own strong architectural statement in the center of that menagerie. The menagerie at Schönbrunn, now in the Viennese suburbs, built for Empress Maria Theresa in 1752 by her husband, Holy Roman Emperor Francis I, also has a Rococo pavilion decorated with sculptures of cherubs and animals as its centerpiece of twelve radial paddocks, and is the best known surviving example of this zoo style.

MENAGERIES IN ENGLAND

In contrast to the developments in Europe, and especially the intellectual approach of the French, the menageries in England at this time were crude and rough affairs. Traveling menageries had first appeared in England at the turn of the eighteenth century, trundling along the muddy lanes of Britain with wagons full of assorted wild animals, shuttered up so that passersby might not get a glimpse of something for nothing. With the advent of the new turnpike roads, the speed of travel, the size of the collections, and the consequent popularity of these menageries greatly increased. An advance agent distributed handbills boasting of the animals' ferocity and strength. The town would be plastered with posters a week before the show arrived: "This is to give Notice, to all Gentlemen, Ladies and others, that there is to be seen in the town of Ludlow, a Curious Collection of Living wild creatures. . . ." The menagerie, comprising sometimes as many as thirty wagons, was set up on three sides of the town square and covered with sheets of canvas. A wooden façade, gaudily painted with larger-than-life wild beasts locked in mortal combat, was erected and the show was ready for the audience.

These traveling animal shows ranged in size but the largest was George Wombwell's. At Bartholomew Fair in 1840 he displayed a puma, jaguar, serval, and an ocelot; leopards, genets, hyenas, wolves, jackals, coatis, and raccoons; a polar bear, a sloth bear, and black and brown bears; kinkajous, porcupines, three elephants, an Indian rhinoceros, gnus, zebus, axis deer, monkeys, birds, reptiles, three giraffes, a "white antelope," a "black tiger," and a "river cow," presumably a hippopotamus (Keeling, 1984).

The largest of the fixed menageries in England was in London. Pidcock's Exhibition of Wild Beasts, founded late in the seventeenth century by Gilbert Pidcock, who had originally managed the collection as a traveling menagerie, was housed in a commercial building in the Strand, approximately where the Strand Palace Hotel now stands. It was a notorious establishment, described by one contemporary as a "disgusting receptacle," and the subject of many complaints, including the frightening of horses in the street by lions roaring in tiny cages within. Later it changed its name and ownership, becoming Polito's Royal Menagerie, but conditions remained as dreadful. Cages for solitary animals such as lion, tiger, hyena, various monkey species, an ostrich, and even a rhinoceros (billed as a unicorn) and an elephant, were so small that the animals could barely stand or turn around. The conditions were very similar to those that exist today in the zoos of economically emerging countries such as China. The menagerie collection included a tapir, which was taken for walks in the neighboring streets, and a degenerate mandrill named Happy Jerry, who drank gin, smoked a pipe, and once had dinner (venison) at Windsor Castle with King George IV (Broderip, 1847). Less happy had been the fate of one of the other animals: a celebrated and very large Indian elephant named Chunee. Released from the tight confines of his cage each night and walked along the Strand, he was understandably a celebrated resident of London and once achieved theatrical fame in the stage production *Blue Beard* at Covent Garden Opera House. One day in 1826 Chunee went berserk, suffering from a diseased tusk. After three days of massive cage battering, a hastily assembled file of soldiers was brought in to kill the poor beast. They fired no fewer than 252 musket balls into his body before running out of ammunition. Chunee then succumbed to a harpoon thrust. For many years, Chunee's skeleton, his skull riddled with bullet scars, was a feature at the museum of the College of Surgeons, until destroyed in 1941 in a bombing raid (Brightwell, 1952).

Dead elephants, however, can still draw crowds and generate income for a crafty showman. When the elephant in George Wombwell's menagerie died en route to a fairground, a rival posted banners declaring that he had "the only live elephant in the fair." Wombwell attracted even larger crowds with notices proclaiming "the only *dead* elephant in the fair." It was

Wombwell, too, who sought crowds by staging at Warwick a fight between a lion and six mastiffs. Some say the dogs ran away, and some say Wombwell called off the fight when he perceived his lion was in danger. But whatever happened, it clearly tells something about the conditions and attitudes that prevailed.

NATURAL HISTORY AND THE DEVELOPMENT OF CHARACTER

The emergent Industrial Revolution in England was viewed as evidence that scientific methods and instruction could propel humankind into a future of unbounded progress. This was not just wishful thinking by industrialists and investors. Eighteenth- and nineteenth-century intellectuals and social reformers in Britain were convinced theirs was an age of unbridled potential. Alfred Russel Wallace, co-originator of the Theory of Evolution, held passionate beliefs, strongly influenced by fellow Welshman Robert Owen, that the new industrial society would lead to an idealized Earthly paradise. A common opinion began to grow that a world of freedom and happiness was about to transcend a history of adversity and ignorance. The spirit of improvement that fueled the technological innovations of this age created a middle class addicted to self-improvement through the acquisition of enlightening information. Owen established the Institute for the Formation of Character, which had the world's first day nursery and evening classes for workers. Pursuit of knowledge came to be generally seen as a measure of temperament, demonstrating a capacity to be productive and resourceful with one's leisure hours. Attendance at drinking houses, dogfights, fairgrounds, and gambling salons was believed by the new middle classes to encourage debauchery and idleness. Recreation was to be used primarily for self-advancement.

Subscription libraries, newsrooms, and discussion groups flourished. Literary and philosophical societies abounded for the families of industrious and respectable businessmen. Most popular of all were the various natural history societies, botanical and horticultural clubs, and the geological societies. Private menageries had already become highly desirable features of aristocratic gardens in seventeenth- and eighteenth-century England as ever more exotic species were shipped there in vessels filled with plunder from all parts of the world. Explorers and naturalists such as Alfred Russel Wallace, Henry Bates, and Richard Spruce were able to finance themselves by supplying museums and wealthy collectors with specimens stripped from the Amazon rain forests. Middle-class, educated families soon began collecting natural history specimens on such a scale it became a mania. Natural history books were amazingly popular. The Reverend J. G. Wood's *Common Objects of the Countryside* (1858) sold one hundred thousand copies in one week. The comte de Buffon's forty-five-volume *Histoire Naturelle* was an essential accouterment to Europe's most elegant drawing rooms. In 1763 the *Critical Review* announced that natural history had become "the favorite study of the times."

The appeal of gathering shells, fungi, butterflies, beetles, birds' eggs, fossils, rocks, and wildflowers and sorting them into classified collections was that the whole family could enjoy it together, closely examining the wonders of Nature at home. Natural history studies offered a solid basis for emotional, aesthetic, and intellectual satisfaction for middle-class parents and their children. They also served to reveal God's handiwork and, thus, a path to understanding divine wisdom.

Not everyone perceived this benefit. As natural history collecting became more fashionable, it inevitably became more of a contest. Ladies and gentlemen of wealth sought to amass the greatest and rarest collections, and there were many scandals of theft, bribery, and intimidation. Those who could afford to collect and maintain living collections advanced to the top of the competitive heap. Aviaries for exotic birds and hothouses for tropical plants became the vogue. New technologies of forced hot-water heating and of cast iron ribs to hold panels of glass meant that those with sufficient means could now build very large glasshouses. The duke of Devonshire's cost £30,000, the equivalent of hundreds of gardeners' annual salaries. Designed by Joseph Paxton and Decimus Burton for the duke's country estate, Chatsworth, this conservatory was three hundred feet long and sixty-seven feet high, and used eight coal-fired furnaces to generate hot water for seven miles of piping. Lit by twelve thousand gas lamps and sporting a fountain reaching fifty feet high, this impressive private hothouse was open gratis to the public.

Some of these lofty jungle habitats were built as public projects. They had special appeal for northern Europeans, cut off from sunshine and greenery for several months each year. Glazed winter gardens were built in capital cities and resorts, early heralds of tourist theme parks. Hector Horeau constructed one in Paris that was six hundred feet long: an Arcadian people's palace decorated with two hundred thousand camellias, palm trees sixty feet tall, orchestras, restaurants, bowling alleys, dance floors, swaths of lawn, perfumed fountains, and gas lamps to assist the moonlight falling through the glass.

More than a century before the Jungle World exhibit was built at the Bronx Zoo, Andrew Jackson Downing, writing in the *Horticulturist* in 1854, envisioned a public greenhouse in New York "where the people could luxuriate in groves of the palms and spice trees of the tropics." Joseph Paxton, co-designer of the duke of Devonshire's hothouse, created the extraordinary Crystal Palace in London's Hyde Park, for the Great Exhibition of 1851. He hoped it would be transformed into a winter garden, "supplied [with] the climate of southern Italy, where multitudes might ride, walk, or recline amidst groves of fragrant trees and leisurely examine the works of Nature . . . regardless of the biting east winds or the drifting snow" (Kohlmaier and von Sartory, 1986). Similar motivation encouraged his unfulfilled dream of a Great Victorian Way; a glass-enclosed roadway 72 feet wide and 108 feet high, enclosing houses, offices, and hotels, winding through nine miles of London, adorned with palms instead of plane trees.

The technical ability to construct very large greenhouses was essential to the development of exhibition buildings in botanical gardens. These big structures, however, were almost completely ignored in zoos until recent years. The reason is interesting, and perhaps it is because tropical plants refuse to live in enclosed structures without abundant light, whereas animals from the same regions, unfortunately, can survive in gloomy confines. They had to tolerate dank and dreary cages simply because, like the human dwellers of slums, they could.

But the Industrial Revolution brought more than new building techniques. More progressive laws for animal welfare and protection began to appear in the nineteenth century. Greater sensitivity toward wild animals had emerged in the seventeenth century as utilitarian views of Nature had started to give way to more moral judgments; coupled with the strong sentimentalism of the nineteenth century, these attitudes now gained force. The Royal Society for the Prevention of Cruelty to Animals (RSPCA), the first of its kind in the world, was founded in England in 1824, paving the way for similar societies in France, in 1845, and in the Netherlands, in 1864. (Despite securing a royal appendage, the RSPCA has never managed to dissuade the British royals from their persistent slaughter of animals. Albert, Prince Consort to Queen Victoria, enthusiastically promoted fox hunting; Edward VII shot wild animals in abundance on his world travels; George V shot dead more than a thousand birds one short December day and killed more than twenty tigers and ten rhinos on a two-week trip to Nepal. Hunting and shooting wild animals continues as a traditional pastime for the Windsors.) These new societies were based on humanistic ideals, aimed as much at elevating humankind as protecting animals. A proposal for a parliamentary committee to investigate bearbaiting and dogfighting, for example, was intended to determine the extent to which such activities corrupted public morals.

Wild birds began to elicit special concerns and, probably because they can fly, became favorite symbols of freedom. Some people objected to the growing fashion for private aviaries. More vigorous arguments arose against the extravagant waste of wild bird feathers by milliners, against bird nesting by egg collectors, and against bird hunters. Ornithologists started to turn from examining birds along the sights of a gun barrel to using telescopes and, later, cameras.

Despite this burgeoning interest in wildlife, there was as yet no great zoo in England. That was about to change. A new zoological garden, which was to be built in London early in the nineteenth century, would set a new standard and a new style of zoological garden development for the world. It also established a new level of popularity for such ventures.

The immensity of London Zoo's early success was due partly to its novel concept, its landscape-design qualities, and its scientific approach to zoo management. Most especially, however, a confluence of many different societal attitudes, standards, perceptions, and beliefs that had been growing and changing during the past two centuries or more converged in the perfect time and place.

These shifts and modulations formed a strange cluster. There were the wide and intense interest in natural history, recognition of the value of scientific studies, the satisfaction of order in the Linnean system of classifying plants and animals, a new acceptance of wild animals as objects of interest, and a paternalistic concern for animal welfare. The recent and passionate interest in wild places was fueled by the mass distribution of adventure stories about strange and distant lands and the continual discovery of previously unknown exotic animal species. In addition, English society had developed a desire for new public parks in its expanding cities and warmly embraced the evolution of the informal English-garden style. It also supported the notion of wholesome family outdoor entertainment. These changes, combined with the growth of a comfortably well-off middle class, and the emergent belief in edification as the mark of a civilized and progressive society, all combined to set the stage for the wild success of what was in essence the world's first zoological garden.

THE ZOOLOGICAL SOCIETY OF LONDON

The 1820s in Europe were an inspiring time. French novelist Théophile Gautier (1869) wrote in his memoirs, "All was young, new, exotically colored." It was the beginning of the Romantic era: an era of rapid change and growth and the beginning of the shift of wealth and responsibility from the aristocracy to the bourgeoisie. And it was in this atmosphere that Sir Stamford Raffles initiated active discussions in scientific circles on the feasibility of establishing a new zoological park in London.

Raffles, discoverer of the world's largest flower, *Rafflesia,* and founder of the Port of Singapore, had been an avid collector of animals during his travels in the Far East. He had written to his cousin in 1825, "I am much interested in establishing a Grand Zoological collection in the Metropolis, with a Society for the introduction of living animals . . . bearing the same relations to zoology as a science that the Horticultural Society does to Botany. We expect to have 20,000 subscribers at £2 each" (Vevers, 1976). The 1825 prospectus for the society observed that "It has long been a note of deep regret to the cultivators of Natural History that we possess no great scientific establishments either for teaching or elucidating zoology; and no public menageries or collections of living animals where their nature, properties and habits may be studied. In almost every other part of Europe, except in the metropolis of the British Empire, something of this kind exists. It would well become Britain to offer to the population of her metropolis . . . animals brought from every part of the globe to be applied either to some useful purpose, or as objects of scientific research."

Things moved quickly. The Zoological Society of London's first council met 5 May 1826 and approved the preparing of plans by architect Decimus Burton; the gardens opened in Regent's Park less than two years later. Despite this speed, Raffles did not live to see his dream materialize; he died of cerebral hemorrhage, age 45, in July 1826.

Although there had been initial concern about placing the zoo in such a remote spot as Regent's Park, the Zoological Gardens met with immediate popularity and soon became the most fashionable venue in London. When a nineteenth-century music hall song, *Walking in the Zoo on Sunday,* became a hit, the world had its first "zoo" and in 1867 the new word appeared in the Oxford English Dictionary.

During its first hundred years, London Zoo introduced many other innovations, including the first reptile house in 1849, the first public aquarium in 1853, and the first insect house in 1881. But its greatest influences were the scientific principles on which it was founded and its setting in a large, public, open park with informal, naturalistic landscaping. This was to be the pattern of development for an eruption of new zoos all over the world through the entire nineteenth century. No other zoo development of similar significance occurred until the twentieth century, when Carl Hagenbeck opened his new zoo in Hamburg in 1907, in which, for the first time, large wild animals were presented in full-scale illusions of their wild habitats.

London Zoo was the first to have a philosophy of acquisition, and there was from the beginning active debate on the composition of its collection. One faction of the Zoological Society, mainly landowners, wanted to emphasize animals such as eland, ducks, pheasants, and trout, with potential for acclimatization and subsequent breeding for farms and the parks of manor houses, while the other faction, principally naturalists, wanted a collection based on taxonomic diversity. Each faction agreed, however, that they did not want to follow "the vulgarity and sensationalism" of the Exeter

'Change menagerie. The land-owning gentry, however, came to realize that they could fill their needs via agricultural societies. They also noted that a breeding farm set up by the Zoological Society at Kingston had quickly failed, due to weak demand for stud fees of zebu and zebra. Thus the scientific view prevailed, and soon the zoo was progressing toward the curators' "stamp collecting" mentality of trying to get full sets of each genus. A Tibetan wild horse acquired in 1859 was valued because it completed the zoo's "series of wild species of the genus Equus" (Zuckerman, 1980). One can appreciate the value of having all representatives of any collection especially for comparative purposes, but the logic seems questionable when using such criteria for selecting living wild animals for display. The dilemma hinges on whether the principal purpose for maintaining the zoo collection is scientific research or public education.

The planning at London's new zoo was based on a philosophical foundation of scientific advancement and didactic enlightenment. It was quite different in appearance, mood, and direction from the prevailing French style. The Royal Botanic Gardens, in the London suburb of Kew, had opened sixty years earlier and pioneered the concept of setting out a collection of plants based on Linnean methodology (Hill, 1915). This system had become associated with all that was progressive in natural history. Even though it assumed a preordained plan of creation, it was considered the embodiment of reason and of rational thinking and was integral to the concept of first the botanical and then the zoological garden.

During the Age of Enlightenment, "truth" was to be made absolute and attainable by the application of science and scientific methods. As knowledge grew, truth would inevitably be more clearly revealed. Thus, it was reasoned, society must of necessity make progress, moving forward not just materially but into a future of unbounded perfection. Once nature was understood and all its secrets revealed, a rational world could emerge in which society would also arrange itself harmoniously. Science, intelligence, truth, and progress were viewed as inseparable. Botanical gardens and zoological collections were an essential component in the public dissemination of these new understandings.

Botanical gardens in the eighteenth century were thus conceived as scientific landscapes, offering natural history instruction within a picturesque setting. The careful arrangement of the collections in taxonomic groupings was designed to encourage observation and learning. London Zoo, following Kew Gardens's impetus, consciously based its layout on taxonomic presentation of its animal collection: the first for a zoological park. Taxonomy arranges animals and plants in related groups. Thus, the animal kingdom is first divided into vertebrates (animals with backbones) and invertebrates (which comprise at least 95 percent of all known animal species). These are then grouped into classes. The vertebrate classes are fish, amphibians, reptiles, birds, and mammals. These classes are then subdivided into orders, nineteen for the mammals. The *Carnivora,* with about 250 living representatives, is one example of a mammalian order. Each order is further divided into families, such as the *Felidae,* with its thirty-six species of cats, which is split again, into genera, such as the genus *Panthera* for the big cats, and then into species: the lion is *Panthera leo.* London Zoo used this system for the layout of the exhibits in its new gardens. All the big cats were grouped in one building, and all the other carnivores, such as canids, bears, and mustelids, were located in their own groups in other places. Similarly, the *Perissodactyla,* or odd-toed hoofed mammals, were displayed separately from the *Artiodactyla,* their even-toed cousins. Primates were housed in one building and subdivided by families such as Old World monkeys, lemurs, or apes. Birds were similarly split into their taxonomic groups, with different display areas and buildings for parrots, penguins, *Falconiformes* (the so-called birds of prey: eagles, hawks, and vultures), cranes, owls, passerines, and so on.

Good modern zoos maintain detailed records of daily happenings, and the first such Daily Report prepared at London Zoo, and maybe the first ever of its kind, was written on 25 February 1828. It noted that the otter had died "in consequence of a diseased tail" and that the emu had laid her fourth egg. The house for llamas was "in progress" and the "boundary wall for supporting the bank next to Bears Pit began."

By April the Zoological Gardens were ready for opening, with about two hundred species in the collection. By late summer it was clear that the venture was a huge success, even though admission was initially restricted to members of the Zoological Society and their guests (although tickets could be bought under the counter in local pubs). The *Mirror,* 6 September 1828, recorded that "The grounds are daily filled with fashionable company,

The tradition of setting animals on display in taxonomic groupings, introduced at the London Zoo in the early nineteenth century, has proven extremely tenacious, probably because it appeals to scientific logic and the strong sense of catalogued order that prevails among collectors. That it is boring and generally meaningless to visitors never seems to have bothered curators. This example of displaying the family Rhinocerotidae in the order Perissodactyla is from the Berlin Zoo in the 1960s. The sterility of the enclosures underscores the related obsession with neatness. (Photo courtesy of Berlin Zoo.)

notwithstanding the great migrations which usually take place at this season of the year and almost depopulate the Western Hemisphere of fashion. The Gardens, independent of their zoological attractions, are a delightful promenade, being laid out with great taste and boasting a beautiful display of flowers. The animals, too, are seen to much greater advantage than when shut up in a menagerie, and have the luxury of fresh air, instead of unwholesome respiration in a room." In 1830 William IV presented the animals from his royal collection at the Windsor Park menagerie to the Zoological

Society, and in 1832 the royal collection in the menagerie at the Tower of London was also relocated to London Zoo. Tigers, lions, an Asian elephant, a "zebra of the plains," a llama, blackbuck, various small carnivores, a "New Holland emu," alligator, boa, anaconda, crane, rodents, and more than one hundred rattlesnakes were taken from the dank and dark confines of the Tower and moved to Regent's Park (Brightwell, 1936), now engaged in an almost perpetual program of improvements and expansion, and made all the more popular by these regal acknowledgments.

Promenading in the park was laden with far more symbolism than we might today imagine. A group of reform-minded Parliamentarians had formed a Select Committee on Public Walks and in 1833 compiled a report advocating construction of public parks as a means of promoting a more healthy, respectable society. "Notice the advantage which public walks (properly regulated and open to the middle and humbler classes) give to the improvement in the cleanliness, neatness and personal appearance of those who frequent them. A man walking out with his family among his neighbors of different ranks will naturally be desirous to be properly clothed" (Wyburn, 1966). The report gained national recognition and support in a society that correlated outdoor exercise, moral discipline, acquisition of knowledge, cleanliness, and godliness.

Middle-class families visiting the zoo had the opportunity to simultaneously witness their social superiors in public, imitate their manners, and learn of the wonders of nature. That all this happened in a built environment, orderly and categorized, made it all the more palatable. It was a setting in which tamed Nature gave respectable English families a place to spend recreational time in the productive pursuit of greater cognition and self-improvement.

The *Daily Telegraph* of 4 January 1870 noted that "The Zoological Gardens are simply the most popular exhibition in London. We all go to the British Museum for instruction's sake, but we visit the Zoological Gardens for amusement as well as instruction." London Zoo typified the modern zoo in that it started as a scientific enterprise, but never escaped a fascination for the sensational. William Makepeace Thackeray caught the flavor of the place in his *Sights of London* (1850), saying that he liked to often visit the zoo, "keeping away from the large beasts, such as the bears, who I fancy may jump from their poles upon certain unoffending Christians; and the howling tigers and lions, who are continually biting the keepers' heads off" (Vevers, 1976).

The extent of the zoo's popularity is difficult to imagine today. After the royal family, it was probably the most publicized institution in nineteenth-century Britain. Many of its animals became national pets, cherished by the media and the public and eliciting great public compassion. New arrivals often roused a mania. These rashes of public interest began with the arrival of a quagga in 1831; an Indian rhinoceros in 1834; four giraffes in 1836; and the first orangutan, Jane, in 1837. (In 1842, Queen Victoria found her successor, Jenny, to be "frightful, and painfully and disagreeably human" [Raby, 1996]). Eighteen thirty-five had been the year of the chimpanzee. Tommy had arrived at Bristol and was taken from there to London by night coach, after great difficulty by his keeper in persuading any carriage proprietor to accept such a passenger. Theodore Hook, playwright and lampooner, summed up the mood of London that summer with a popular ditty that began, "The folks in town are nearly wild / To go and see the monkey-child." It was, however, the advent in 1850 of Obaysch, named after the island in the Nile where he was captured, and touted as the first hippopotamus in Britain for half a million years, that caused the wildest furor. All London society flocked to see the animal after his arrival on 26 May, appalled and delighted by the monstrosity of his form. The immensely popular British author Thomas Babington Macaulay wrote, "I have seen the Hippo both asleep and awake, and I can assure you that, asleep or awake, he is the ugliest of the works of God" (Blunt, 1976). It was just the thing to help swell attendance, which more than doubled that year to more than 300,000 visitors.

(The Greek historian Herodotus named the *Hippo potamios,* meaning "river horse," though this name and the description that he gave suggests he may never have seen or heard one. The hippo, he said, "has the mane and tail of a horse . . . and a voice like a horse's neigh." Pliny, the prolific Roman documentor of natural history, added that when the hippo leaves the water to graze, it always walks backwards, to confuse any hunters or pursuers [Krutch, 1961]. A later travel writer, the intrepid Mary Kingsley, could not determine whether the hippo was Nature's first bungled attempt at creation or a final exhausted fling: "Here just put these other viscera into big bags—I can't be bothered any more" [Blunt, 1976].)

Royalty by royalty. Obaysch, dubbed by the press as H. R. H. (His Rolling Hulk), sleeps unconcernedly as he is photographed at London's Zoological Gardens in 1852 by Don Juan Carlos María Isidro de Borbón y Braganza, Count of Montizon, grandson of Spain's King Carlos IV. For many years the zoo was considered to be one of the most socially acceptable venues in class-conscious London. (Photo courtesy The Royal Archives © 1999 Her Majesty Queen Elizabeth II.)

Firsts piled one on top of another. Three years after the hysteria of the hippopotamus's arrival, the zoo opened the first public aquarium. In 1853 the *Literary Gazette* described it as "an elegant aquatic vivarium . . . a light airy building sixty by twenty feet in area, containing around its transparent walls fourteen six-feet tanks of plate glass. . . . They enclose masses of rocks, sand, gravel, corallines, seaweeds, and sea-water; and are abundantly stocked. . . . The whole in a state of natural restlessness, now quiescent, now eating and being eaten."

The world's first reptile house was opened at the zoo in 1849. The world's first insect house followed in 1881, "constructed of iron and glass on three sides, with a brick back to it" as the Annual Report for that year prosaically described it. Indeed, much of the early architecture at London Zoo was decidedly utilitarian, rooted more in basic functionality than the ornamental garden styles of the day. Decimus Burton's giraffe house of 1836 was no more than a garden shed, albeit taller than any other, but in no way exotic in style or adornment, which is puzzling, considering the novelty and enthusiasm that was washing around the zoological gardens.

THINGS BEGIN TO DECLINE

From 1859 to 1878, a period of great importance for the zoo's growth and a time when English architects such as Norman Shaw and Philip Webb were experimenting with new directions, achieving lightness and animation in their picturesque buildings, the society, notes Peter Guillery (1993) in his intelligent appraisal of the buildings of London Zoo, unfortunately relied upon the design talents of Anthony Salvin. A mediocre architect at best, he was a Fellow of the Zoological Society who successfully courted the board's favors during an era of the zoo's rapid expansion. Salvin designed many major buildings, including the eastern aviary, the lion house, the antelope house, the monkey house, the parrot house, and the elephant house, which historian Simon Schama (1995) likens to "a sort of rustic almshouse for pachyderms." Even Salvin's biographer describes him as a "dismal talent," and his designs for the zoo are depressingly unimaginative, clumsy attempts of the suburban-cottage *orné* style. Even more damaging was his influence on the zoo's overall layout. Burton's delightfully informal and irregular plans were steadily replaced with the dull practicality of straight and sensible broad paths, a suitable stage for Salvin's pedestrian buildings. Gradually the zoological gardens were deromanticized, tamed into orderly domesticity.

Nothing improved after Salvin's retirement. From 1882 to 1903, building works at London Zoo were under the direction of an engineer, Charles Trollope, and the exhibition buildings he added to the grounds were equally undistinguished. Later, under the governance of Sir Peter Chalmers Mitchell, the zoo paid closer attention to the quality of its architects, employing very talented designers such as John James Joass and Berthold Lubetkin. This progressive attitude was reignited in the 1960s, when special effort was made to employ the very best of the day. Sir Hugh Casson produced the marvelously expressionistic if admittedly not very functional elephant house, and Anthony Armstrong-Jones (later Lord Snowdon) and Cedric Price designed the clever and still impressive northern aviary, built on (and seeming to float over) a very awkward site.

Always, however, the emphasis was on seeking architectonic rather than naturalistic design solutions. Perhaps this was an inevitable direction for a perceived national zoological garden in an urban park in the British Empire's premier metropolis. It is peculiar, though, that many of London Zoo's exhibit buildings resembled nothing so much as provincial railway stations or modestly decorated factory sheds. As decade upon decade saw more and more architectural development, the result, no matter how fine an occasional individual piece might be, became increasingly oppressive and visually monotonous. The presence of wild animals in the park was ever more at odds with the uninspired and excessively overbuilt environment.

Even in recent times, in the heart of the zoo and adjacent to the main entrance, an awful metal shed housed monkeys and apes, including a famous gorilla, Guy, who survived almost thirty years there in solitary confinement. This building was replaced in the early 1970s with the much larger Sobell Pavilions for apes and monkeys, designed by staff architect John Toovey and described by him as "providing the maximum involvement of animal and visitor with landscape" (Zuckerman, 1976). Unfortunately, its slabs of brown brick, steel space-framing, metal climbing furniture, and naked concrete combine to create a sense of depressing sterility. Views into the monkey dens, when not obscured by reflections from the glass, reveal rooms more reminiscent of an impoverished police cell than anything that celebrates the soft green natural world of wild primates. The animals appear as nothing more than sad incongruities. The central location and very large area of the Sobell Pavilions seem to give official sanction to this form of brutal design solution at London Zoo.

Too many somber and dreary buildings accumulated over a hundred and fifty years or so have created a generally depressing atmosphere at London

Zoo. A stultifyingly large number of its built forms are now on heritage conservation lists, frustrating any possibility for the zoo ever to remake itself. Poor attention to visitors, evidenced by atrocious food services, and an inflexible adherence to past practices such as trying to maintain too many big animals in inadequately small spaces led to such dissatisfaction that by the 1980s attendance had plummeted to record low figures. The zoo that had set many benchmarks had ossified. Media attacks had been becoming ever more bristly since the late 1960s and not just in the popular press. In the 1970s the *Ecologist* opened an editorial with the statement, "The London Zoo is a shameful establishment where wild animals [live] in totally inappropriate conditions." A Sunday newspaper magazine profiled the zoo as a "Beasts' Belsen," after Bergen-Belsen, the Nazi concentration camp.

Not only was the zoo's public face looking decrepit, but its scientific credentials were being questioned. The society's secretary, Lord Zuckerman, was a scientist who held a powerful position in society but never established the London Zoo as a particularly credible scientific institution. He founded the zoo's Institute for Comparative Physiology in 1962 and the Institute of Comparative Medicine in 1964, but these large new institutions were not relevant in any useful way to London Zoo's operation and did nothing to contribute to the well-being of the animals. When cuts had to be made in new exhibit buildings, an invariable hazard in any construction program, they were typically made at the expense of the animals' needs. The size of the outdoor paddocks at the new elephant house, for instance, was greatly diminished, over the objections of curator of mammals, Desmond Morris, rather than curtailing, for example, the excessive height of the building. The multitalented Morris, with a keen interest in animal behavior, had the vision and energy to lead the zoo into a progressive new era, but was too often thwarted by a stuffy and megalomaniac administration. In 1966 he left the zoo in frustration. It was a significant loss at a critical period. Later, there was further erosion, this time of institutional credibility, when Zuckerman embarrassed his fellow zoo colleagues with his closing address to the 1976 World Conference on Breeding Endangered Species in Captivity. At a time when zoo directors Gerald Durrell and William Conway were pleading for better and more vigorous attention to captive breeding, Zuckerman moved in the opposite direction. He publicly questioned the need to give priorities

By the 1970s, the London Zoo was looking decrepit, after more than a century of building depressingly dreary exhibits. The only change to the ugly concrete mass of the Mappin Terraces in fifty years was the addition of jagged shards of slate along the edge of the bear moats. (Author photo.)

to wildlife conservation. During his long tenure as secretary of the Zoological Society of London, during which time he was given first a knighthood, then a lordship, at the behest of his friend, Prince Philip, Zuckerman presided over the steady decline of the London Zoo until it was on the verge of collapse.

A new team took over at the zoo in the early 1990s, with goals to improve visitor services, offer lively education programs, and pursue a new dedication to conservation, but unfortunately, a board with a long tradition of micro-management still hamstrings the professionals. London Zoo set many initial benchmarks for zoos, but in the 1980s it was such a depressing place, so abandoned by the public, it seemed it must fail and close, whereby it could have set a surprisingly novel and useful trend for British zoos. With some 260 licensed zoos on its small islands, Britain has far too many, and generally of such mediocrity, that it would be better if 90 percent of them were closed. Those British zoos such as Marwell, Howletts, and Jersey, which have conservation as a main goal, and the zoos at places such as Bristol and Chester, and, of course, London Zoo and its country estate at Whipsnade, could then receive the attention and support they deserve, rather than resources being dissipated among hundreds of second- and third-rate facilities.

CHAPTER FOUR

ROMANTICISTS AND MODERNISTS

Throughout the nineteenth century, following the success of the new zoo in Regent's Park, numerous cities around the world, but particularly throughout Europe, were busily constructing new zoos. Bands played and flags flew for grand-opening days all across the Continent in major cities almost every year of the century. Most of them followed the pattern of London Zoo, basing their new institutions on scientific principles, setting out their collections in taxonomic groupings, involving themselves in research, and placing pedagogic labels on the exhibits.

The most adventurous zoo architecture, too, appeared in Europe. Many zoo buildings of this period were vastly expensive, even sumptuous, but it is nonetheless clear that the zoo managers and designers had little knowledge about their animals. They knew virtually nothing about the wild habitats of the animals or of their natural diets, their breeding habits, natural groupings, or lifestyles. Bread and milk mixed with boiled rice was the staple diet for numerous species in most zoos. The daily ration for the elephant at the Jardin des Plantes was eighty pounds of bread, twelve pints of wine, and two bucketsful of gruel. It was not uncommon for keepers to feed apes humanlike dinners, from roast beef and potatoes to cream and cake. Fruit was for some reason often considered generally unsuitable for monkeys.

With no solid information, the zoo designers of the age used architectural exotica and looked to the animals' legendary histories or countries of origin for inspiration. Mosques and minarets were commonplace. Antwerp Zoo in 1846 built a precisely detailed replica of an ancient Egyptian temple for its elephant and giraffe house, complete with authentic hieroglyphics. The ostrich house at Cologne Zoo resembled a Hindu place of prayer.

One could have expected and hoped to see truly revolutionary developments in French zoos, but the history of zoos in France has been principally a story of lost opportunities, greed, and squalor. The small area of the Jardin des Plantes had soon become crowded after its foundation as a zoo site in the early years of the Revolution, and in 1860 Isidore Geoffroy Saint-Hilaire prepared an annex in the Bois de Vincennes, hoping that conditions there would be "more favorable to the fecundity" of the animals. Unable to raise support for that notion, he then courted La Société Zoologique d'Acclimatation and was thus able to secure fifty acres in the Bois de Boulogne for a Jardin Zoologique d'Acclimatation.

The Rothschilds, as well as many other notable families, and the emperor all enthusiastically supported the new zoo. These founders invested one million francs because part of the zoo's plan was to domesticate, breed, and sell new animals for agriculture and industry. Thus the collection included Angora goats, llamas, alpacas, silkworms, a herd of forty cows from which the zoo sold one hundred glasses of milk a day, cormorant fishing displays, a trout farm, and a horse-riding school. The zoological exhibits were awful and included a blind brown bear, kept in a pit by himself. Equally revolting was Monsieur Martin's Mechanical System for Fattening Fowls: a massive rotating cylinder holding tiers of boxes, each containing a chicken that was force fed by a seated attendant operating a foot-operated piston pump connected to a rubber tube shoved into the bird's throat. The chickens were available for sale to visitors, as were most of the animals on display.

Following in the footsteps of grandfather Étienne and father Isidore, Albert Geoffroy Saint-Hilaire became a zoo director, in 1865. It was bad timing. The Jardin Zoologique experienced its first financial loss that year. Officials then closed it during the Franco-Prussian War of 1870–71, when Paris was under siege. All the animals at the Jardin des Plantes and at the Jardin Zoologique, including lions, bears, a giraffe, two elephants, and a hippopotamus, were slaughtered by order of the authorities and handed over to the butchers' shops. Zoo officials had made attempts earlier to send the animals to the safety of Antwerp Zoo, but the evacuation was brought to a halt within five days when all trains stopped running. The authorities took over the grounds of the Jardin Zoologique to house the 130,000 sheep and 20,000 cattle required to feed a starving city. After the war, things became even worse. In the anarchy and insurrection of the Commune that followed, Paris was under siege for the second time in three months. The zoo was this time in the middle of the battle and for nearly two months steadily disintegrated under the bombardment of shells and bullets.

When it opened again, in 1874, the Jardin Zoologique faced huge mortgage debts, had lost much of its infrastructure and, worse, was no longer regarded as a fashionable place. In desperation, the director tried new ways to attract audiences, displaying caravans of Nubians, Eskimo families, Argentinean gauchos, and a troupe of dwarves that were touring Europe (as the "Kingdom of Lilliput") and offering music concerts, fairs, puppet shows, and light opera (Osborne, 1996).

FANTASTIC ARCHITECTURE

France has never produced any innovative public zoos nor, for some reason, have any of the southern European countries. The best have always been in northern latitudes, and the grandest and most prolific of the nineteenth and early twentieth centuries, in every way, were Germany's. They built superb structures, gathered the most impressive collections, and set the highest standards of the day. Of all the German zoological gardens, the most impressive architecture was in the Berlin Zoo.

It began when Professor Martin Lichtenstein, a physician who, after living in South Africa for many years, had become director of Berlin's Museum of Natural History, visited London Zoo in 1832 and determined to build a zoo in Berlin. In 1841 he obtained the approval of Friedrich Wilhelm IV to set up a planning committee, which chose a site in the former royal hunting grounds of the Tiergarten. It also hired architect Peter Lenné, one of the professor's long-time supporters, to prepare a plan. His idea was to keep the place principally as a park, with very few buildings or enclosures. When the zoo opened in 1844 it had a bear castle, monkey house, bird house, bison paddock, and a restaurant (Heilborn, 1929).

Soon, other German zoos were appearing—by 1870 in Frankfurt, Cologne, Dresden, Hamburg, and Stuttgart—with expansive and elaborate exhibition buildings and gardens full of flowers. Berlin's zoo by comparison began to look inadequate, even shoddy, and officials began developing plans for a complete change.

A new director, Heinrich Bodinus, hired in 1869, laid out the new exhibits on a taxonomic basis and set himself to gather as large and diverse a collection as possible. Architecturally, his most spectacular decision was to build the new exhibits in fantastic and exotic styles. Thus, the zoo acquired a carnivore castle, antelope mosque, pachyderm temple, Moorish bird house, and a monkey palace, each one a carefully detailed and elegantly proportioned structure.

Bodinus's successors, Maximilian Schmidt and Ludwig Heck, continued his tradition of exotica and, like him, consistently employed architects of great quality. Heinrich Kayser and Karl von Grossheim designed an outstanding Japanese-style building for water birds in 1897 and an Egyptian temple in 1901, for ostriches, closely modeled on the giraffe house built in Antwerp in 1856. Scholars from Berlin University ensured the decorations and hieroglyphics were accurate. In 1905, architects Zaar and Vahl produced a heavily decorated Russian timber manor house for European bison. Most wonderful of all were the works of Hermann Ende and Wilhelm Böckmann, who produced for Berlin Zoo some of the most superb zoo architecture the world has ever seen, including the delightfully airy monkey house of 1884, the sumptuously detailed antelope house that resembled a mosque, and an extraordinary elephant house in the style of a Burmese temple in 1873 (Klös, 1969). (By 1939 Berlin Zoo had the largest and most important animal collection in the world, with about 4,000 mammals and birds of almost 1,500 species and 8,300 reptiles, amphibians, and fish of more than 750 species. The bombing raids and street battles at the end of World War II saw the almost-total destruction of Berlin Zoo. Only 91 animals survived.)

Other architects in zoos all over Europe let their imagination flow in fanciful romanticism. They built replicas of castles, Tudor cottages, copies of ancient Greek temples, Swiss chalets, Renaissance pavilions, and whimsical follies. They built in gothic, rustic, classical, Chinese, Indian, and any other style they could conceive. Ignorant of what the animals truly needed, these nineteenth-century designers built hundreds of fantastic new zoo buildings for animals taken from deserts and forests, savannas and tundra, but with no attempt to replicate the natural homes of the occupants, many of which were in spaces no larger or better than those in the old menageries, with social species typically enduring solitary and brief lives.

RADICAL IDEAS

An extraordinary exception to such attitudes was the revolutionary proposal by American zoologist Robert Garner, who studied apes in zoos and in the wild and prepared recommendations for housing gorillas and chimpanzees that were astounding. Like many others in the late eighteenth and early nineteenth centuries with intense interest in the origins of language, Garner was fascinated by the possibility that apes used a language and claimed that he had learned many words of chimpanzee, "but two of them are much greater in volume than it is possible for the human lungs to reach." Ironically, it is the apes that are unable to produce the sounds required for human speech, not necessarily because they lack the mental machinery, but due to their larynx being positioned too low in the throat. (The same situation exists in human infants up to the age of three or four months, which is why they can breathe and suckle simultaneously.) If it weren't for the location of their voice boxes, chimpanzees might have been able to learn to talk to Garner. Certainly apes have demonstrated the capacity for nonverbal communication with humans and can comprehend complex and novel spoken commands. Garner wanted to discover how apes communicate in the wild, and on 9 July 1892 he set sail for Africa to observe gorillas and chimpanzees in their native habitat. He took with him an extraordinary device—a collapsible wire mesh cage, not to enclose captured animals, but to live in while he made his observations.

He arrived at his destination, Lake Ferran Vaz, two hundred miles up the Ogowe river in the French Congo, after several months of travel, and in April 1893 he erected his cage base in the jungle, naming it Fort Gorilla. Painted dingy green, to blend in with the surroundings, the cage was fitted with canvas roof covers, a canvas hammock, a folding chair, and a board hanging from wires as a table. He had a kerosene stove, blankets, a camera, tins of food, and a pet young chimpanzee, Moses, purportedly fond of corned beef and sardines. "In this novel hermitage I remained most of the time for one hundred and twelve days and nights." This was a minuscule period of time compared to the extraordinary number of years that Jane Goodall has dedicated to research on wild chimpanzees, but Garner nonetheless learned much that was entirely new about gorillas and chimpanzees, and it was so different from the tall tales carried home by white hunters about savage and blood-thirsty monsters that he knew many of his findings would be discredited. He regretted, he said, "that it devolves upon me to controvert many of the stories told about the great apes, but finding no germ of truth in some of them, I cannot evade the duty of denying them."

Sitting alone for hours and days in his cage, Garner absorbed the moods and sounds and essence of the jungle. He described its gloom and stillness, the clearly varying layers of vegetation, the heat and humidity, the sometimes "endless, voiceless solitude," and the various animals he encountered, including the elusive gorilla. From this experience, Garner developed firm ideas about the conditions that they required in captivity and he published his findings in 1896. If these had been followed, years of misery and deprivation for these highly intelligent animals would have been avoided in many zoos.

All apes in zoos, said Garner, should have an enclosure at least fifteen feet high. The south side and half the roof should be of glass panels, with canvas covers to regulate the sunshine. In summer the enclosure should be left completely open, to admit fresh air and rain. Garner was strongly of the opinion that "apes need not be pampered. Half of the gorillas that have ever been in captivity have died from overnursing." The enclosure must, he declared, have a "sandy loam or vegetable earth" floor, and there should be a wide but shallow pool, filled with "a dense crop of water plants." A steam coil in the pool base was to keep the water warm, and a rain spray of tepid water was to be activated once or twice a day for at least one hour at a time. Away from the pool was to be "a strong tree, either dead or alive," for climbing and heaps of dead leaves for the animals' comfort and play. Noting that loneliness was a "fruitful cause of death," he stated that apes must always be in the company of their own kind, to entertain and amuse themselves;

"otherwise they become despondent and gloomy." He was equally adamant that no visitor be allowed on any terms to give them any kind of food. These considerations may now seem logical, but nothing remotely like this was then in existence, and many of his most fundamental proposals did not begin to be implemented for more than seventy years.

A contemporary of Garner's, working at a zoo in San Francisco, also revealed attitudes far ahead of his time, with good intuitive instincts about the needs of captive animals. These are recorded in Ernest Seton-Thompson's study of the animals at the new national zoo in Washington, D.C. "It has long been known," he said, "that caged animals . . . suffer from a variety of mental diseases," and that most zoo animals had a life expectancy of only three years; but he also noted some remarkably progressive management techniques at Robert Woodward's Gardens in San Francisco where a Mr. Ohnimus, superintendent of the zoo, had "kept most of his animals . . . alive, healthy, and happy . . . [for] sixteen years." (Sadly, a local minister launched a vitriolic campaign against this establishment, claiming that the noises of the animals disrupted his Sunday services. The authorities closed the gardens and ordered that the animals be shot in their cages.)

Ohnimus told Seton-Thompson that the great secret of his successes with animals was caring for their *minds* as well as their bodies. He explained that in captive conditions typical of the day, "several species commonly end their cage lives in lunacy. . . . Captive bears are apt to fall into a sort of sullen despondency. Foxes and Cats often go crazy. . . . The higher Apes and Baboons rarely thrive in cages. Soon or late they become abnormally vicious or else have a complete physical breakdown. . . . Their bellies are well cared for, but few keepers have learned that in each animal is a mentality, large or small, that ought to be considered." Ohnimus's techniques were effectively simple. For example, he varied the feeding times for the animals in his care and adjusted the amount and frequency of feeding according to the animal's natural eating patterns. He offered variety and novelty by adding or removing cage furnishings and occasionally moving animals from cage to cage. In particular, he made time each day to play with the animals. This was remarkably humane and enlightened management for its time, but in spite of Seton-Thompson's publication that gave the profession the opportunity to benefit from these experiences, the techniques were almost universally ignored. Similar animal-keeping practices would not become commonplace for almost another hundred years.

Robert Garner's proposals were ignored, too, dismissed as too radical and, some said, too expensive, although that had not seemed to be a problem with the vast number of opulent zoo buildings of the nineteenth century. Officials believed that putting gorillas out in the open air, giving them rain sprays, and telling the public that they could not feed the animals was a frightening risk. Constructing large and impressive buildings, however, brought only praise and prestige. Ornate and massive zoo structures had become national status symbols resulting in intense competition among zoos to create ever more extravagant monuments. They competed with each other in two principal ways: to have the largest collections of different species and to have the most majestic architecture. Zoo designers therefore continued to search for inspiration in pattern books, history journals, culture guides, and architectural magazines. No one expected to find satisfaction examining the dangerous and unsavory wild homes and habits of the animals. Indeed, the zoo animals were perceived as being distant from the wild in more than just miles of distance. It was almost as if zoo managers believed the zoo animals to be different from their counterparts in the wild. Perhaps when observed in refined and elegant environments, treated as pets, and fed basically human diets, the animals appeared to be not wild but civilized, raised to a higher level.

But a triumphant departure from this approach occurred in a major new German zoo, and it came not from any professional architect or member of the zoological scientific community, but from an animal trainer and collector. In the twentieth century's first decade, Carl Hagenbeck made a great leap from the traditions of the past century and instantly created a new paradigm for zoo design. It was a romantic vision, but based on images of natural habitats, not of other zoos, and was developed with respect and care for the well-being of the animals.

HAGENBECK'S NEW DIRECTION

The zoological park in Hamburg, Carl Hagenbeck's Tierpark, is unique in many ways. The first zoo in the world to combine naturalistic landscapes,

The palatial interior of the elephant house at the Berlin Zoo, designed by Ende & Böckman in the 1860s. Elaborately detailed and decorated, it served its civic purposes admirably, proudly demonstrating the grand status of the new zoological gardens. The sterility and amount of space for the animals, however, were no better than the conditions that had prevailed in the wretched menageries of the past. (Courtesy of the Berlin Zoo.)

barless enclosures, and groups of mixed species, it is also privately owned by one family and operated today by the sixth generation of Hagenbecks. It all began modestly, even accidentally with Carl Hagenbeck Sr.'s fishmonger business in Hamburg. He had a contract to purchase the total catch from certain fishermen, so when in 1848 they took him six seals that had become entangled in the fishermen's nets, he was obliged to buy the animals. To recoup his money, he exhibited the seals at a local fairground. Encouraged by this profitable venture, he became more involved in animal dealing for zoos and traveling menageries; in 1863 Hagenbeck abandoned the fish business and purchased Gotthold Jamrach's nearby menagerie, which was housed in a barn, eighty feet by thirty feet, with stalls for carnivores on one side and herbivores on the other, and boas and pythons in crates down the middle. A courtyard at the back housed birds and monkeys, an African rhinoceros, and the first Sumatran rhinoceros seen in Europe. Carl Hagenbeck Jr. inherited this menagerie in 1866 and built it up to become the world's largest wild animal dealership. He moved to a larger, two-acre site in 1874 and opened his first Tierpark (which has since become the German term for a zoological park), with lion, elephant, monkey, and reptile houses and a birds-of-prey aviary. His collection grew to be one of the most valuable in existence, containing gerenuks, pigmy hippos, manatees, and Mongolian wild horses; by the 1880s it was attracting huge crowds (Reichenbach, 1996).

The idea of building a different type of zoo began to form in his mind. In his autobiography (1910), Hagenbeck says, "I desired, above all things, to give the animals the maximum of liberty. I wished to exhibit them not as captives, confined within narrow spaces, and looked at between iron bars, but as free to wander from place to place within as large a limit as possible, and with no bars to obstruct the view and serve as a reminder of captivity."

It was important to Hagenbeck that he demonstrate to scientists as well as to the public that many tropical animals could acclimate to live outdoors in temperate climates. Hermann Landois, founder of Münster Zoo, had already experimented successfully with this notion, but most zoos were unwilling to take the risk with their tropical animals (Ehrlinger, 1990). Hagenbeck was convinced the animals would be healthier in the fresh air; many zoo professionals ridiculed the idea. He also fervently wished to demonstrate the benefits of giving generous space to the animals and to display them in an environment that was natural for them.

After years of diligently searching for a suitable site, he bought an estate in the Hamburg suburb of Stellingen and transformed twenty-five acres of flat potato fields containing six trees into a landscape of mountains, gorges, lakes, forests, and islands. To build his zoological park, Hagenbeck hired a wide range of skilled artisans. Most importantly, he employed Urs Eggenschwyler, a Swiss sculptor who had created the first barless or moated enclosures for lions in his own small private zoo in Zurich. Eggenschwyler, having lost his hearing as a youngster and suffering badly at school as a result, had early in life developed a great interest in animals as the living beings that he could best relate to. He turned to sculpture as a medium in which he could eloquently express himself, and his feelings for animals kept in small cages were in close sympathy with Hagenbeck's. They made a formidable team. Eggenschwyler made three-dimensional models for the new

Carl Hagenbeck in the company of trained lions at his Tierpark in Hamburg. He introduced radical methods of training animals, using kindness and conditioning rather than the harsh techniques typical of the period. (Photo courtesy of The Hagenbeck Archive.)

zoo, sculpting the landscape with rock formations and gorges based on sketches from real geological formations.

Hagenbeck experimented with novel ways to determine the height and width of the moats that were to keep people and animals safely separated. He investigated carefully the animals' capacity both at the high jump and the long jump: "My method was to take a stuffed pigeon and fix it to a projecting branch of a tree . . . [and] then let loose in turn lions, tigers, and panthers." Based on the results from these experiments, he "considered it safe to surround the carnivore gorge at Stellingen with a trench twenty-eight feet wide" (Hagenbeck, 1910).

The new zoo, initially comprising two panoramas—Africa and the Arctic—opened in May 1907. It was an immediate and great success with the public. Their excitement of seeing lions unobscured by bars was so great

One of the great panoramas at Hagenbeck's new zoo, in the Hamburg suburbs, with huge rock formations designed by Urs Eggenschwyler. Zoo exhibits of this grandeur and scale had never been attempted before. Hagenbeck wanted not only to show wild animals in representative habitats, but also to prove that animals from other regions could live healthy lives in large open enclosures. A critical factor in the concept of *landscape immersion* is that animals and people be submerged within as authentic a replica of the natural habitat as possible. Hagenbeck came tantalizingly close to discovering this idea. His panoramas were surprisingly realistic considering how novel they were, but visitors were still very much on the outside of the scene, looking into a stage set. (Photo courtesy of The Hagenbeck Archive.)

that Hagenbeck found it prudent to control the crowds by charging an additional fee to approach the moat edge. Hagenbeck's professional colleagues, however, did not approve of the new park. He had initiated some of the most important innovations in zoo design, the first to display animals associated by regions and the first to develop barless panoramas, but, as Edward H. Bean, the innovative director of the Chicago Zoological Park, noted, it was "much ridiculed by his contemporaries" (Livingston, 1974). Ludwig Heck, director of the Berlin Zoo, particularly abhorred the new style, claiming that it would threaten the taxonomic, scientific approach to zoo planning (Baetens, 1995). Such closed-minded criticisms would unfortunately continue for some time. Critics made very similar negative observations about the new concepts of bioclimatic zoning and landscape immersion when they were introduced at Woodland Park Zoo, Seattle, in the 1970s. Many zoo curators and directors thought the extra landscaping of public areas in those concepts to be superfluous, and others strongly disliked the move away from taxonomic grouping. Hagenbeck's zoo, however, set new standards in visitor satisfaction and in animal welfare that were rarely improved upon for well over half a century.

The panoramas, for which Hagenbeck had received a patent in 1896, were made up of a series of enclosures, laid out like theater stages, each one behind and slightly higher than the other and separated by hidden moats. Artificial rockwork and plantings concealed the holding quarters and service ways. The hidden moats were an adaptation of the English pastoral ha-ha—sunken fences or ditches that prevented cattle and sheep in landscaped parkland from encroaching upon the flower gardens but kept them in the overall view of the general landscape. The obscured moats, dramatic rockscapes, and numerous ponds and lakes created scenes of expanding vistas in the most audacious zoo development to that time. The African panorama was the first to generate the illusion of an open savanna, populated with gazelles, flamingos, storks, cranes, antelopes, zebras, lions, and, in the distance, ibexes and wild sheep on rocky outcrops. Moreover, the scale of these panoramas was breathtaking.

Around the world, zoos tried to copy Hagenbeck's designs, which in many cases was a mistake. Rather than studying natural habitats and examining geological formations to understand what caused their particular shapes and colors, other zoos merely attempted to mimic what Hagenbeck and Eggenschwyler had created, and they in turn were copied by other zoos. At each step in this generally reductionist process, the rockwork tended to become ever more a degraded caricature of the beauty and fascination of real rock formations. At its worst, and sadly most common, this resulted in mere heaps of stones cemented together, with little similarity remaining to the original structures. This problem persists in zoos, especially in Europe; many designers still prefer to copy ideas from other zoos than to seek inspiration from studying natural habitats.

Zoos around the world copied Hagenbeck's panoramas, but rarely with his sense of grandeur and naturalism. The lumpen forms of the Mappin Terraces at the London Zoo, built in 1914, are some of the worst large-scale examples that copied the forms but not the spirit of his work. The strange surrealism of the terraces was still evident when this photograph was taken in the 1970s. (Photo © Zoological Society of London.)

Another mistake was that other zoos did not approach the problem with the same conviction and avidity as Hagenbeck did. London Zoo's Mappin Terraces, built in 1914, were an early effort to reproduce something of similar scale and style, but the results were merely lumpish and drab. Other than its large size, the structure had little else to commend it. The Parisian zoo of Vincennes, opened in 1931, was also created on the Hagenbeck mold but lacks the bold drama of its mentor. Zoos in the United States showed a somewhat better understanding of the grandness of scale, and in the 1930s, when the Works Progress Administration funded hundreds of projects to renovate zoos across the country, produced several exhibits that had distinct artistic merit. Denver Zoo's mountain habitat built in 1918 was the first example of Hagenbeck's style in America, and it even surpassed his level of accuracy by taking plaster molds of rock formations in the Hogback Range west of the city. St. Louis built large, open bear grottos in 1921 that also went to further levels of authenticity than Hagenbeck's artists and sculptors, making direct copies from limestone bluffs and giving careful attention to the natural stratification and geological form of the rockwork. Houston also produced a good early example with its sea lion pool, and San Antonio Zoo developed some exceptional bear and monkey exhibits in the Hagenbeck style in the 1920s. Detroit Zoo employed the Hagenbeck family directly to design many of their exhibits, built between 1928 and 1938 (Austin, 1974). No one, however, planned an entire zoo and tackled the overall problem with such verve as Carl Hagenbeck, and none matched his overall boldness of scale.

RAINING ROCKS

Most zoos were unwilling or unable to aim for the high standards that Hagenbeck defined. It is a great irony that the creative breakthrough he made and the high visual drama he introduced to zoo design resulted in a plague of mediocrity and boredom. As more zoos copied Hagenbeck's style from each other with no real understanding of its philosophies and standards, they started to look more and more alike. Fake rockwork, ranging from average to far below that, pervaded the zoo scene. Grottos, islands, cliffs, and fanciful imitation rock formations with no logical or geological

The Bear Mountain exhibit under construction at the Denver Zoo, built under the direction of zoo superintendent Victor Borchert in 1918, was strongly influenced by Hagenbeck's work. It was the first naturalistic zoo exhibit in the United States. Casts made in the Hogback Range west of Denver were used as molds for the concrete structure. The final structure was colored with dyes and conifers were planted on top. The results, though impressive, did not please everyone. William Hornaday, director of the Bronx Zoo, complained that "some half-baked western zoos" had succumbed to "the Hagenbeck fad." He was particularly concerned that zoo visitors were too distant from the animals in such enclosures, and grumbled that the cost of artificial rockwork was an unnecessary expense. Very similar complaints are still voiced by some zoo professionals in opposition to naturalistic exhibition habitats. (Photo courtesy of the Denver Public Library, Western History Collection.)

Ugly environments are astonishingly well accepted in zoos. This example from the Philadelphia Zoo in the late 1980s is typical of numerous other "monkey islands" in zoos worldwide. (Author photo.)

name sprang up in zoos as the abiding solution to exhibit design. By 1933 landscape architect Richard Addison was noting that "in the world of zoos it has been raining rocks. Big rocks, little rocks, rocks of all shapes, sizes and hues."

Artificial rockwork continues to be the ubiquitous solution of many designers to most zoo design problems and too often is a mere pastiche of the real thing. Much of it appears insubstantial or looks to be exactly what it is, poured mounds of concrete, rather than any natural formation. Even when sufficient care is taken to use real rock molds, they are too often put together to make overall shapes that have no basis in reality. The unremitting expanses

At a zoo that should know better, the bear exhibit in the mountain habitat at the Arizona-Sonora Desert Museum, designed by the Potomac Group, is formed of generic concrete rockwork, unnatural in form and detail. Visual insult is intensified with squared holes in the supposed rock and the very obvious presence of steel doors and bars in the face of the viewer. It is rather difficult to engender a sense of wild habitat in such circumstances, and any chance to interpret the complexities of mountain habitats is wasted. (Photo courtesy of Kenneth Stockton.)

of geometric rock formations at Milwaukee Zoo and Columbus Zoo are a synthetic simplification carrying the subliminal message that any concrete formwork can represent natural habitat. Over the years San Diego Zoo has sprayed vast quantities of concrete to create a multitude of vaguely amorphous and deathly sterile grottos in which to display its animals. The mountain habitat designed by the Potomac Group for the Arizona-Sonora Desert Museum, in Tucson, is formed with artificial rockwork that has no texture that can be found in nature. More damning, and typical of much zoo design, its forms reveal no sensible clue for their existence at that particular place.

When very small, isolated canyons form in nature, as they sometimes but rarely do, there is some specific geological reason for their existence. Zoos, however, reproduce this eccentricity in abundance, as a convenient barrier to keep people and animals apart. Invariably the designers then construct a curving rock wall with an overhang at its top, to enclose the exhibit space. It is a combination that probably does not exist anywhere in nature, yet is ubiquitous in zoos. The example at the Arizona-Sonora Desert Museum further degrades its chances of authenticity by placing, as zoos are wont to do, iron-barred service doors in square holes cut into the fake rockwork and to add insult locates them directly in front of the viewer's eye.

The resultant experience of this "mountain habitat" is a walk past a repetitious series of open grottos, each of the same general size, each of the same general level of interest, and each containing one large carnivorous mammal species. There is no revelation of what a mountain habitat truly is, how mountains are formed in nature, why one finds different types of formations and colors in montane environments, or what a rich complexity of life forms can be found within such a habitat. The saddest aspect is that this poorly designed structure is built at a zoo with a reputation for high standards of realism.

NEW FASHIONS, NEW STYLES

Early in the twentieth century, a shift occurred in people's attitudes toward design, diet, health, and exercise as they began casting away the appurtenances of the Victorian era. The wealthy spent weekends at sanatoriums, the middle class began participating in exercise sessions, and everyone seemed to become convinced of the benefit of exposing their bodies to the rays of the sun.

Zoos of the time were no different. Throughout the Victorian era almost all the new zoos had been built in temperate, mostly northern, latitudes. They had been obsessed with ways to heat their buildings, determined to keep their fragile and expensively obtained animals from the rigors of cold and damp. The Victorians were convinced that disease was transmitted through bad air, and the public spaces of heated animal houses were a special concern. Criticisms of "stifling and ill-smelling conditions," as recorded by the *Illustrated London News* in June 1869, were common.

In the new century, attitudes began to shift. The Modern Age began to associate progress with everything that was clean, open, bright, and streamlined. Architects sought functional expression in their buildings, uncluttered by ornament. Objects from automobiles to cocktail shakers were chromed and made smooth. Housewives languishing in the cells of suburban isolation sought inspiration in magazines that exhorted domestic cleanliness and Formica coffee tables. Science and technology began to produce such wonders as artificial sunlight from quartz incandescent globes. When London Zoo opened its 1927 reptile house, it was the most technologically advanced building of its type in the world, incorporating electric heating with separate thermostatic controls for each individual cage and artificial ultraviolet lights as a substitute for sunshine.

Other zoo buildings reflected the importance of modernity. The Tecton Group of architects, in their designs for zoos in London and Dudley, rejected Hagenbeck's naturalism. The principal in this firm was Berthold Lubetkin. As a modernist-functionalist, he wanted to demonstrate his design commitment to animal health empirically rather than impressionistically, and he sought to reveal science and technology with an architectural vocabulary and a rationalist approach to zoo design.

The *Architect and Building News* of 1 June 1934 criticized Hagenbeck's panoramic approach to the theatricality of zoo design, complaining that "it allowed the very shy animals to hide themselves from the public gaze, almost indefinitely." That was not going to be a problem in the Tecton Group's new gorilla house for London Zoo, the first building ever designed by the firm and described by a contemporary critic as marking "the last word in modern architectural efficiency" (Guillery, 1993). Floor surfaces were designed so that it was more comfortable for the animals to sit near the viewing window than at the back of their room. (This is not a unique approach to pleasing visitors; Paignton Zoo in England, which describes itself as a "tropical paradise," uses grids of metal points protruding from the otherwise barren concrete floor to dissuade animals from sitting where the zoo director

Modern architects delighted in the new style of zoo architecture developed by the Tecton Group in the 1930s, shown here in the form of the elephant house at the Whipsnade Zoo. This may surprise us; the place looks very much like a prison or a torture chamber. But modern (and postmodern) architecture tends to value form more than function, with much more attention given to superficial style than to the well-being of a building's occupants. (Reproduced courtesy of the *Architectural Review.*)

does not want them to sit.) Instead of the crude vagaries of outside temperatures, the gorillas had the benefit of modern air-conditioning, which, said the zoo's Annual Report for 1932, "washes the air free from fog, dust, and germs." An especially novel feature was a retractable wall and roof, to allow London's rare sunshine into the space, but it was technically awkward, rarely used, and never tried again at the zoo.

Lubetkin's approach to zoo architecture was no less theatrical in contrivance than Hagenbeck's. His design for London Zoo's penguin pool was a severely minimalist Antarctica, rather than an impressionistic version in the Hagenbeck style. It expresses nature functionally and rationally; it is concrete poetry, playing with the plastic possibilities of what was then a new material. As abstract sculpture it received rave reviews from the architecture profession, who made much of its "elegance and technical virtuosity," and the Annual Report for 1934 noted that "The Society has been congratulated as a pioneer in artistic and practiced architecture." To the aesthetically inclined, the penguin pool was a symbol of the liberating possibilities of the brave new age of modernism. Sir Charles Reilly, a professor of architecture, was so captivated by its "perfect unity" that he wrote to the *Architect's Journal* to express the wish that he could have just such a place for his own abode and hoped that he might "live long enough to have a small town house, I suppose with one ramp for my wife and another for myself. . . . No doubt I shall have to simplify my habits before I am worthy to live in such a thing of beauty, but that would be very good for me as for most of us" (Reilly, 1934). No one asked the penguins how they enjoyed having to simplify their needs to live each day in their minimalist pit.

This manifestation of scientific enlightenment as architecture heralded a depressing phase of zoo history. It was the start of the Disinfectant Era. "Modern" meant efficient and hygienic. Thus animals from forests and deserts, evolved over millennia for life in complex and environmentally dense habitats, were now to live in zoo exhibits designed principally for water hoses. From the late thirties through the sixties (and beyond that date in many instances) most major zoos adopted a pose of scientific purity. The result was clinically sterile cages with walls lined in glazed tiles, usually white or pale green, smooth concrete floors, and cage furnishings reduced to a stainless steel pole and a cantilevered slab.

In exhibits of this era, plate glass denied even audio contact with the public. Keeper access was always via a steel door, which slammed shut against steel flanges, the noise reverberating at a painful level in the hard acoustics of the cages. The animals had nothing to interact with other than their feces or food. Nothing inside these cages ever changed from one day or one year to the next. Every part of the animal's environment was controlled and immovable. Keepers were forbidden to vary cleaning routines or to introduce novelty food items into the diets. In the most pseudo-scientifically progressive zoos, such as Philadelphia, the animal diets were reduced to the convenience of prefabricated vitaminized biscuits, nutritionally sound but sensually defunct. The boredom and monotony of this regime are beyond comprehension to any human who has not spent years in solitary confinement. No wonder the animals came to resemble the minimalist standards of their care—with calcium rich bones, glossy coats, and minds numbed into oblivion.

As is often the case, one rare individual was advocating a different approach, but was out of sync with the rest of the zoo world. Edward H. Bean, director of Chicago's Brookfield Zoo, was already explaining in 1929 that the common approach of systematic arrangements for zoo collections was "not entirely satisfactory" and produced monotony. Bean was too far ahead of his time to be heard, just as his advice about artificial rockwork was ignored. He had recommended that open exhibits should be produced first as a scale model and "be shown for criticism to geologists" (Livingston, 1974). This advice is still ignored today, despite its obvious and sound sense.

Another puzzle is why the general acceptance of sterile environments, which created such boredom for wild animals in zoos, was repeated in similar environments for the paying public. The fashion at this time was to lay out cages in long straight rows, circles, or some simple geometric shape, and thereby ensure that visitors soon experienced mental as well as physical fatigue from walking past one glass-fronted sterile cage after another in obsessively tidy and tedious arrangements. It is a problem that has bedeviled all sorts of museums, and it stems from the unfortunate fact that many museum professionals, be they zoologists, Egyptologists, botanists, or aquarists, have the mindset of collectors. They yearn for full sets of things—be they snakes, sarcophagi, cycads, or cichlids—and want to arrange them in order. This undoubtedly has value for researchers, especially for comparative studies, but it has very limited application for exhibitry. The usual result for viewers is boredom and fatigue.

When taxonomic collections comprise living animals, a special sadness accompanies the tedium. Sets of wild animals in sterile cages, as if they were no more than bundles of feathers, piles of fur, or cuts of meat, have blighted the history of virtually every zoo in existence. As recently as 1968, the American Association of Zoological Parks and Aquariums (AAZPA) selected the layout of an exhibit building at Staten Island Zoo as "an excellent example" of zoo planning. Very similar to a cross-shaped Gothic cathedral in layout, this building had thirty reptile cages in rows against the walls, a double row of twenty-four small cages for mammals, and an identical arrangement for birds. Rows of fish tanks filled another area. It was a perfect formula for monotony, but the Zoo Association liked it because it "systematically arranged display sections for the four major vertebrate animal classes . . . within a single building" (Curtis, 1968). The association made no comment about the inevitable boredom of the viewer or the diminished quarters in which the animals had to exist.

HEDIGER'S LEGACY

In 1950 Heini Hediger, then director of Basel Zoo in Switzerland and later, after an altercation with its board, director of Zurich Zoo, published his book *Wild Animals in Captivity: An Outline of the Biology of Zoological Gardens.* No architect or zoo manager since then can be justified in not knowing how to care for zoo animals and properly design for their needs. Five years later Hediger published his *Studies of the Psychology and Behaviour of Animals in Zoos and Circuses.* In these books he gave cogent arguments for and clear examples of a *biological* approach to zoo design and animal care. He explained the concepts of territory and rank within groups and the phenomena of flight distance. He examined the importance of play and the value of using natural materials and feeding natural diets as well as ways to avoid stereotypic behaviors of bored animals. In particular, Hediger made it clear that *quality* of space, not mere quantity, was the critical factor in meeting wild animals' needs in captivity.

Zoo designers could never again plead ignorance as an excuse for keeping arboreal animals, like orangutans, on bare floor exhibit spaces, for feeding monotonous diets at set daily feeding times, keeping the animals separated from natural vegetation, denying animals the ability to hide from view of potential enemies, or keeping social animals in solitary confinement. Yet these ubiquitous zoo practices continued unabated. Hediger's careful studies of animal behavior and his beautifully logical reasonings should have been welcomed and adopted by his professional colleagues, but his words remained those of an unheard voice in the unseen wilderness, ignored for decades by zoo managers and designers around the world.

Animals in the wild live in an environment of great complexity, with much spatial and temporal variation. Until recent years most zoo environments were by comparison sterile, and care regimes were predictable and unchanging. These deprived conditions, no one should be surprised, cause extreme tedium resulting in many severe problems, from stereotypic and repetitious behaviors to aberrant sexuality, and from hyperactivity to almost total inertia. The conditions can affect even the most basic behaviors. When a zoo-raised tiger in the 1970s was released into a large outdoor area at the World Wildlife Safari in Winston, Oregon, it began to stumble and walk so erratically it was thought to be ill. The tiger, which had known nothing but the flat floor of a zoo cage its whole life, was having great difficulty coping with a natural substrate with some variation in its terrain (Hutchins, Hancocks, and Crockett, 1978).

A myth has become established that successful breeding in captivity is somehow proof that animals are being kept in the right conditions. Many zoos persist in perpetuating this story, but in aberrant environments animals can become hypersexual, a signal of mental distress rather than a healthy sexual function. Pregnancy is not always a sign of a healthy and mutual sexual relationship. Proof is more likely to be indicated when the mother carries the fetus to full term and then successfully raises her baby with natural maternal care and passes on to it normal patterns of behavior for its species. But in too many instances, baby zoo animals are immediately taken from their mother for hand raising. When a gorilla gave birth one afternoon at Woodland Park Zoo in Seattle in the early autumn of 1983, among the tall grasses beneath a shady tree in the exhibition habitat, the media were deliberately kept uninformed in order for the mother to remain undisturbed. Though kept under close observation, Nina was allowed to care for the infant herself, in her own quiet way. It was a memorable sight. A teenage female gorilla sat in close and rapt attention, carefully watching the proceedings and occasionally reaching out to gently touch the new baby. Two adult male gorillas sat apart, some distance away, equally carefully *not* watching, pretending to be preoccupied with the distant view or occasionally closely peering at a leaf or a piece of grass, but constantly taking surreptitious sidelong glances at the birth scene.

The story of this unusual birth barely made mention in the local press. At about the same time, a gorilla was born at a Midwest zoo; it was taken from the mother and rushed to its new home in the zoo nursery with a small army of reporters and video cameras in close flight. That story received national television coverage. For such attention, zoos will loudly publicize the arrival of new babies. Many of them will publicly display their baby animals in specially built zoo nurseries. These babies have usually either been taken from their mothers or, as a result of abnormal environments, have been abandoned by them.

If all this seems inexplicable, how can one account for the fact that the biological and environmental needs of animals in captivity have historically received scant attention? People like Hediger, the zoo biologist, Ohnimus, the zookeeper, and Eggenschwyler, the zoo artist, have always been in the minority. Their soft approach in caring for animals has traditionally been the exceptional, not the typical. It should surprise no one that the majority of people who have worked in zoos did so because they liked them. It is rare for people who dislike zoos, and want to change zoos, to choose to work in them. In too many instances motivation for those attracted to the zoo profession have been desires for control, prestige, and ownership. It is not unusual, and is frighteningly revealing, to hear zoo managers talk of "my" elephants—or gorillas or tigers or whatever. There is enormous prestige, as beguiling to a bureaucrat as it was to pharaohs and princes, in "owning" exotic and powerful beasts.

Fortunately there are signs of a new breed of zoo directors, especially across America. Ron Kagan at Detroit, Terry Maple at Atlanta, Palmer Krantz at South Carolina's Riverbanks Zoo, and Gary Geddes at Washington

State's Northwest Trek Wildlife Park are (relatively) young people who represent a growing number of senior zoo administrators who are not ashamed to reveal a passion and compassion for the animals in their care and a sincere consideration for their well-being. It is not that such individuals have been absent in the past, but the AAZPA (now known as the AZA: American Zoo Association) tended to be somewhat of an Old Boys Club until fairly recent times, reflecting its time and place as a culture that did not expect or value sensitivity. Perhaps the growing number of women in senior positions has had an influence. Barbara Baker in Pittsburgh, Gail Forman in Utica, Patricia Simmons in Akron, and Kathryn Roberts of Minnesota Zoo are examples of major players in the profession who have not felt it necessary to mimic stereotypic male behaviors or attitudes to progress their careers or their institutions. It is also encouraging to see many women now involved in zoo keeping, as well as other young people with university degrees and a love of books, of learning, and of Nature. This shift in the compositional mix of zookeepers must surely bring benefits. Only two decades ago zoos still generally hired zookeepers for their muscular strength. Now the profession is on a much more solid academic base. It is even permissible for keepers to express emotions about the animals in their care, to profess their love for them. In the old days, colleagues looked askance at a keeper like George "Slim" Lewis, who had such remarkable rapport with elephants and who cried over the conditions he would find in zoos and circuses. Zoo keeping still demands physical strength and endurance but has admitted new standards of affection. It is shown in the enthusiasm with which many good keepers today devote their energies and personal time to making the lives of their charges more comfortable and enriched in many creative ways. Scrubbed floors are no longer the measure of the day's job done.

QUICK, CHEAP, AND NATURALISTIC

Hediger (1950) explained that there were two essential and basic ways to increase environmental complexity for zoo animals: spatially, by adding objects of interest and play, and temporally, by making changes from time to time. James Foster, the veterinarian at Seattle's Woodland Park Zoo from

1972 to 1986, with a strong interest in preventive health care, recognized that a zoo veterinarian had to consider the whole animal, including its behavior, personality, and psychological as well as physiological needs. Other staff at Woodland Park Zoo enthusiastically supported his ideas. Shackled by tiny budgets, the staff nevertheless made many dramatic changes to last until they could build larger and more complex habitat-based exhibits.

The changes, initiated in the mid-1970s, were cheap, quick, and simple, but they were unique. For example, a cage for caracal lynx was modified to resemble their natural desert habitat by adding sand, gravel, volcanic rocks, weathered tree branches, and dried sagebrush, all of which the keepers collected at no cost. For a year, the caracals had free access to an adjacent, unmodified cage, where their food was. The animals chose to spend more than 80 percent of their time in the naturalistic enclosure and usually carried their food there to eat it. A similar solution was equally successful for ocelots, except that this cage was lushly planted with donated house plants, such as palms, rubber plants, philodendra, and dracaena, as well as mosses and ferns gathered from the nearby forests.

Keepers added a thick profusion of tree branches to the cages of the old primate house, securing them by flexible joints. This improvement was an apparently simple and obvious thing to do, but it took place in cages that had been devoid of any furnishings other than two fixed metal rods, one vertical and one horizontal, for sixty-six years and were typical of what was to be found in zoos around the world. The keepers also thickly covered the cage floors in the primate house with hay and scattered sunflower seeds and raisins throughout the hay to give the monkeys opportunity for hours of foraging activity.

Shrubs were planted in a pen that had been the home of snowy owls for many years. For the first time the female made a nest, laid eggs, and raised a clutch of babies (later released to the wild). All she had needed was a place to feel secure, out of sight of visitors, behind a small bush. Piles of brushwood and branches placed in a deer enclosure became a focus of activity for the entire herd: newborns bedded down in the brush, adult males used the branches to rub velvet from their antlers and for scent marking, and all of the adults spent hours stripping the bark and eating it. Similarly, large boulders and dead trees were periodically added to bear grottos. Rotting logs and the insects they contained generated much activity and interest for the bears, who usually completely consumed the logs over time.

Similar practices are now more commonplace in zoos and have even acquired an official name: environmental enrichment. Although it has come surprisingly late, it is nonetheless encouraging to see this development and especially the ingenuity and enthusiasm with which zookeepers tend to approach this challenge. It is less comforting to know that some zoo administrators continue to deny the animals this activity. Also worrisome, however, is that others have come to see it as an ultimate goal. They combat criticism of zoos by pointing to their efforts in giving zoo animals the spatial and temporal improvements of environmental enrichment, even though these very techniques and devices are often necessary merely to compensate for the inherent deficiencies of their enclosure designs.

Many zoos dismiss complaints about the extremely unnatural appearance of their environmental enrichment tools as the carping of aesthetic elitists, but it is clear that the physical environment in which a wild animal is displayed has a direct influence upon the perceptions and attitudes of zoo visitors. In 1989 Stephen Kellert, a psychologist at Yale University, and his colleague Julie Dunlap measured visitor attitudes before and after their zoo visits. Zoos with an educational focus and with animals displayed in authentic environments, exerted a positive impact. These effects were greatest for zoos with a high proportion of indigenous wildlife. Visits to more traditional zoos created an increase in negative attitudes to wildlife. There was a measurable intensification in people's indifference to and fear of wild animals after seeing them in an ugly environment. In any case, zoo designers and educators cannot expect people to fully understand what an animal is if it is not presented to them *in context.* A study led by psychologist Ted Finlay (1988) on people's perceptions of animals in various zoo environments and wild habitats clearly demonstrated that the context in which an animal is viewed influences how and what people think of an animal. For the wild animals themselves there is only one appropriate context: their wild habitat. Seen separate from this they become merely oddities.

Landscape architect and zoo designer Jon Coe has said that he judges the effectiveness of a zoo exhibit by "the pulse rate of the zoo-goer." He talks of the need to design exhibits that will make the hair stand on the back of

The naturalistic habitat for lions at the Woodland Park Zoo, Seattle, designed in the early 1980s by landscape architects Jones & Jones. Many traditional zoo visitors intensely disliked the shaggy and wild appearance, complaining that the grass was not mowed as it was at San Diego Zoo and that it was sometimes difficult to see the animals among the foliage. A greater number, however, expressed enthusiasm for such realism. (Author photo.)

your neck (Greene, 1987). Attempts to convince zoos of the need to reach visitors at this emotional level have often been derided as romanticism. Yet if zoos are to help create an informed and aware citizenry that is sympathetic to the increasingly urgent plight of wildlife, they need to be much more effective than they have been in the past. They need to remind their urban visitors of the *wildness* of wild animals. Rather than displaying them as aberrant pets, zoos have to find ways to convince people of the splendor, the beauty, the ruggedness, the *reality* of the wild. A central component of the answer to this problem lies in design and in new ways of presenting animals to urban and suburban audiences increasingly divorced from any daily contact with Nature.

One evening, after attending a meeting at Woodland Park Zoo, Coe was hurrying home through the park, in the gathering dusk. As he passed the lion exhibit, the animals were running and chasing each other through the tall grasses. Coe, who had helped to design this exhibition habitat, paused to watch. More than anyone, he knew all the design criteria that made up this exhibit. He was aware that the animals' space had been elevated above the visitor's eye level to intuitively instill respect, that a hidden moat of provenly effective width safely separated him and the lions, and that the illusion of being in a wild African savanna habitat was the result of careful landscaping and controlled sight lines. Coe was also experienced enough about animal behavior, however, to know that you cannot program for extreme motivation. As he watched the galloping lions, one of the males suddenly paused at the edge of the hidden moat, crouched low, slithered forward, and growled that deep and guttural sound that personifies the primeval. Coe's response was emotional, not rational. The hair stood up on the back of his neck.

CHAPTER FIVE

TOWARD NEW FRONTIERS

It is a little anomalous that the trend for zoological gardens as civic features established itself in Australia several years before it took root in the other New World of America. But until late in the nineteenth century, America was more preoccupied with expanding and exploring its western wild lands than bringing exotic fauna into its new cities. A variety of economic crises and the eruption of the American Civil War also delayed development of such cultural niceties as zoological gardens. Certainly, however, Americans have always been as interested in wildlife as have their antipodean cousins. Hunters returning from the unknown wild lands of the West would often bring equally unknown wild animals, displaying them in taverns or on village greens. The first exotic animal known to have been exhibited in America was a lion, in Boston in 1720, followed a year later in the same city by a camel. A sailor arrived in Philadelphia in August 1727 with another lion, which he exhibited in the city and surrounding towns for eight years.

A menagerie was established in New York as early as 1781, and within seven years had grown to include a tiger, orangutan, sloth, and baboon and crocodiles and snakes. The first elephant was imported from India to America by a ship's captain in 1796 and sold for two thousand dollars. It was first displayed at the corner of New York's Broadway and Beaver Streets and traveled extensively up and down the East Coast, including an appearance at the first commencement at Harvard. Another arrived in Boston in 1804, surviving until 1816. A rhinoceros made its debut in 1831. Menagerie owners were now becoming more competitive, and two American expeditions went to Africa in the 1830s in pursuit of giraffes (Link, 1873).

James and William Howes' New York Menagerie toured New England in 1834 with an elephant, a rhinoceros, a camel, two tigers, a polar bear, and several parrots and monkeys. The following year nine menageries in the United States merged to form the Zoological Institute with a permanent installation in New York City and franchise operations in other principal cities. Their collection boasted a "unicorn," which was actually a one-horned rhinoceros—a deception rather at variance with the institute's stated aim of ensuring that "knowledge of natural history might be more generally diffused and promoted, and rational curiosity gratified" (Flint, 1996).

America's touring menageries slowed to a crawl under the weight of the depression of the 1840s and then to a halt with the outbreak of the Civil War. Only one traveling menagerie of any size existed after the war. The Van Amburgh menagerie traveled the United States for nearly forty years, billing itself, in contrast to the competing circuses, as a "moral show." Fledgling attempts were now underway to develop zoos in some of the larger cities, but even in the late nineteenth century, few of America's zoos had large species such as elephants or lions. At this time, the large collections were on the road, in the traveling menageries.

Unlike their European counterparts, America's menageries and circuses had combined as single traveling shows, with one ticket to see both. This increased the size and the diversity of their collections. There was no shame in their efforts to overpromote these new partnerships or to make grandiloquent claims for their worthiness. Ringling Brothers and Barnum & Bailey's circus menageries advertised their shows as the "World's Greatest Menagerie. 1009 rare animals advantageously displayed in electric-lighted dens where they may be studied at close range in the mammoth Traveling University of Natural History." A note at the bottom promised "scores of cute baby animals to delight the little folks."

When Wombwell's menagerie was auctioned in Edinburgh in 1872, Van Amburgh's agent purchased twenty-two cages of "the largest and best collection of animals ever imported" to the United States. In a three-year period in the 1860s, one animal dealer alone imported twenty lions, twelve elephants, six giraffes, four tigers, eight leopards, eight hyenas, twelve ostriches, and two hippopotamuses for America's circus-menageries. In 1875 Barnum was displaying four elephants, as was his arch rival, the prominent showman Adam Forepaugh, while the circus operated by the Sells Brothers boasted seven. The next year Forepaugh's collection had swollen to twelve, Barnum's to eleven, and Cooper and Bailey's circus owned ten. When Barnum and Bailey merged in 1881, Forepaugh expanded his elephant collection to twenty-five (Flint, 1996). Then the contest expanded into a competition for ownership of the *largest* elephant. Phineas Barnum was to win this battle, with his acquisition from London Zoo of the famed Jumbo, from which he made a fortune, both during the three years Jumbo toured North America, and for some years after, when Barnum displayed his stuffed body.

The boom economy that followed the Civil War encouraged not only the growth of American cities and a shift from agriculture to industry, but, as had happened earlier in an economically expanding England, a growing awareness of loss of wildlife. A greater interest in natural history education also emerged as people's contact with Nature became increasingly fragile. At the same time, a cultural shift saw ever more specialization, the growth of academe, professional management of open spaces into urban parks, the refinement of private collections into public institutions, and, with all this, the early maturing of America's menageries to zoological parks (Stott, 1981).

Government became involved with science, with collecting for the nation, and with funding expeditions and natural history surveys. It eventually came to be seen, in a way never accepted in England, that government should administer zoological collections as well as other cultural institutions for the public good. Thus city, state, and the federal governments have traditionally operated zoos in the United States. "Zoological garden" carries in America, as it has traditionally in Australia and most of northern Europe but never in Britain, the sense of a carefully managed, respectable, scientific institution operationally supported by taxes.

There is no doubt that early American zoos were fueled by a sense of envy that pervaded all views across the Atlantic. European culture, science, and technology were then more impressive and influential. It generated a nationalistic desire not only to equal but to surpass.

AMERICA'S FIRST ZOO

Joel Poinsett, a U.S. statesman, first promoted the idea of a national scientific institution in the United States when he addressed the National Institute for the Promotion of Science in 1841, calling for an institution "having at its command an observatory, a [natural history] museum, [and] a botanical and zoological garden." The Smithsonian Institution, established just five years later, included all these except for, oddly, a zoological collection. The first American effort to establish a zoological park fell to Philadelphia when in 1859 it was granted a charter for "the purchase and collection of living wild animals, for the purpose of public exhibition."

Philadelphia had already founded America's first botanical garden in 1731, as well as its second natural history museum, and was the cultural capital of the early American republic, but there was sparse enthusiasm for raising funds for a zoological park. William Camac, the Zoological Society's first president, complained that "few persons seemed to understand . . . the benefits to be derived" (Segal, 1988). With the outbreak of the Civil War, even the society's interest in a new zoological park was diverted for several years. It was not until 1874, when the gates eventually opened at Fairmount Park, that America had its first public, permanent zoological garden. The collection was large—212 mammals (including lions, zebras, kangaroos, an elephant, a rhinoceros, a tiger, and fifty monkeys) and 674 birds—the buildings impressive and decorative, and the grounds tidy and carefully maintained. But after its first year it did not attract large audiences or enjoy wide appeal. Perhaps the itinerant glamour and hucksterism of the circus-menageries, with their colorful posters and more seductive charms, were too great a competition. Or perhaps it was because the board was trying to run the zoo itself, with an abundance of committees, including committees for animals, accounts, grievances, gardening, entertainment, and employment. Whatever the reasons, there were several years of hard times for America's first zoo, in contrast to the spectacular successes of European countries.

To call Philadelphia's zoo the first in the United States is a point of contention for some zoo historians. Undisputedly it was the first to *plan* a zoological park, with active discussions for such a venture as early as the 1840s, but others note that the Central Park Menagerie in New York, which has since metamorphosed into the Central Park Wildlife Center, actually opened in 1862—more than a decade earlier than the Philadelphia Zoo. Moreover, in 1868 Central Park sent a pair of swans to Chicago, marking the beginning, according to some, of the Lincoln Park Zoo. At the risk of semantic pedantry, however, it seems fair to point out that neither the style, plans, and initial goals of the Central Park Menagerie nor the introduction of two swans to Lincoln Park rates either of them worthy of the term "zoological park."

As with Philadelphia Zoo, a distinctly European (but this time essentially German) approach characterized the early history of Cincinnati Zoo, founded by Andrew Erkenbrecher and opened in 1875. Also like Philadelphia, this zoo experienced great difficulties and setbacks. Oliver Gale's record of Cincinnati's first century, subtitled "One Hundred Years of Trial and Triumph," noted that "some errors were to be expected in an enterprise so novel in America as the establishment of a Zoological Garden, but more have been made than can be accounted for from that reason."

The zoos in both Philadelphia and Cincinnati survived their early faltering days. By 1891 Cincinnati claimed to be the "largest and most complete zoological gardens in the country." Today it has the sad distinction of being the final resting place for the last passenger pigeon and for the last Carolina parakeet, both of which died there in 1914. It also, however, has a record of innovation, having experimented early with barless enclosures. It has continued that progressive tradition under the long directorship of Edward Maruska, with some new landscape-immersion exhibits, an extraordinary collection of rare animals, a strong commitment to research and conservation, and special attention to amphibians and invertebrates, usually neglected in zoos.

In spite of the examples of Philadelphia's and Cincinnati's difficult beginnings, a steady growth of zoos occurred across America. Many of them, like Central Park Zoo and Lincoln Park Zoo, had inauspicious beginnings, often as dumping grounds for unwanted pets or the remnants of bankrupt shows or failed circuses that happened to collapse in that particular town. Thus, the gift of two deer in Buffalo in 1875, the donation of a flock of sheep in 1876 to Baltimore, the bequest of some tame deer to Cleveland in 1882, the abandonment of animals by a circus that slipped out of Detroit one night in 1883, the purchase of the animals from a destitute circus in

Atlanta in 1889, and the donation of a bear cub to Denver in 1896 were the starting points for zoos in those cities.

In similar fashion, the need to find homes for wild animals abandoned after the closure of the Panama-California Exposition of 1915 formed the impetus to establish San Diego Zoo. By contrast, at least two other world expositions had more positive incentives for creating new zoos. The Smithsonian Institution's walk-through aviary and the Hagenbeck exhibits at the 1904 World's Fair motivated the founding of a zoological society in St. Louis, which opened a zoo a few years later. The Chicago Zoological Park in Brookfield was completed in time for the Chicago World's Fair of 1933, taking advantage of the crowds and excitement to advertise its new barless zoo and many rare species (Kisling, 1996).

A NATIONAL REFUGE

As prosperity and boosterism grew during the late nineteenth century, new zoos flourished across America. Just before the end of the century, more than fifty years after Joel Poinsett's proposal and with about one dozen zoos now well established, America acquired its own national zoo in Washington, D.C., set up by Congress "for the advancement of science and the instruction and recreation of the people."

William Hornaday, a young taxidermist who had achieved recognition for his naturalistic, habitat-based dioramas for the U.S. National Museum, had been hired by the Smithsonian Institution in 1882. He soon began planning a diorama of large native animals and in 1886 set off on an expedition to the West in search of bison. What he found there, or more accurately did not find, changed his views and his career and set the early direction of two of America's major zoos.

The expedition spent five hard months to obtain just twenty-five bison, on plains where fewer than ten years earlier countless millions of bison, reputedly the most numerous quadruped the world had ever seen, roamed as they had for hundreds of thousands of years. In 1878 the Northern Pacific Railroad had opened the Missouri region, and a ferocious slaughter of bison exploded with such firepower that when Hornaday arrived, the species was already on the very brink of extinction, falling to an estimated eight hundred animals by 1895. It was seriously considered possible that the elk, pronghorn, moose, bighorn sheep, mountain goat, and mule deer could face the same prospect in the next twenty years. In 1893 the historian Frederick Turner wrote, "The frontier has gone" (Cronon, 1995b). The shock of its passing, so soon after its discovery, prompted the idea of protecting the remaining wild lands.

Hornaday began a campaign to establish "a city of refuge for the vanishing species of the continent," and Samuel Langley, secretary of the Smithsonian Institution, presented a plan to Congress urging approval for a national zoological park (Horowitz, 1996). That America's National Zoo and, later, the New York Zoological Park were founded as responses to concerns about the loss of wildlife has encouraged some to believe that America's zoos were established on a different ethic and for different purposes than European zoos. Cincinnati had started planning a zoo modeled completely on the European concept in the mid-1800s, however, and Philadelphia Zoo, before that, was firmly based on the style of London Zoo. When the National Zoo and the Bronx Zoo opened, they were on very much larger and more rugged sites than any of their European counterparts, but looked remarkably like any major European zoo once they had built some animal houses and dug some flower beds.

Funding approval for the new national zoo did not come easily. Throughout history, kings and princes had set up zoos on a whim. Now a republic faced the difficult task of convincing a majority of its seventy million citizens to invest in one. Debate in Congress was fierce and strenuous, especially about the costs. One congressman, declaring himself a patriot, objected to money being spent for a zoo that would maintain exotic species, claiming this would be contrary to the spirit of the Constitution.

By 1891, however, there was agreement between House and Senate for establishing a national zoo in the Rock Creek area of the District of Columbia. It is of great significance that when appropriation of $200,000 was first approved for the national zoo, it was required to offer refuge so that "native animals . . . threatened with extinction might live and perpetuate their species in peace." No zoo had ever before expressed such goals. Langley immediately hired landscape architect Frederick Law Olmsted, designer of New York's Central Park, with instructions to prepare plans that would maintain the natural quality of the fairly rugged 166-acre site. He was eager to create a different type of zoo. In his annual report of 1889, he described it as a place where "the wild goat, the mountain sheep and their congeners would find the rocky cliffs which are their natural home; the beavers, brooks in which to build their dams; the buffalo, places of seclusion in which to breed and replenish their dying race; [and] aquatic birds and beasts their natural home." Equally as daring, Langley wanted to retain about three-quarters of the site for these purposes, providing only about thirty-seven acres for a public zoo. None of this came about. Congress placed 50 percent of the financial burden for zoo operations on the district taxpayers. Inevitably, the original concept of a wildlife refuge collapsed. An editorial in the Washington *Evening Star* stated that "If the District is to pay for the park as a local institution . . . it wants the entire space thrown open to public use and enjoyment, with none of it reserved for the purely scientific purposes of the government."

The battles and difficult negotiations to found the National Zoo were not followed by much success. Limited funds allowed little to develop on the site. Animals had to be acquired principally by donations, which resulted in odd acquisitions; mainly opossums, raccoons, and various unwanted pets, as well as a circus tiger with advanced mange and a lion abandoned as a cub by a circus and raised in a blacksmith's home by the family cat until it grew to such a size that the neighbors objected. The first bird in the zoo's collection was a discarded pet sulfur-crested cockatoo. A sixteen-page wish list was sent to all U.S. embassies in 1898. It ambitiously included such rare and difficult species as tarsiers, platypuses, koalas, and tuataras.

Adam Forepaugh donated two elephants from his circus and milked maximum publicity from his largesse. He marched the animals through the streets of Washington, the lead animal bearing a placard announcing Forepaugh's gift to the nation. An employee ran on ahead, warning people to get their horses out of the way of the coming procession. At Thomas Circle one weary driver, with an equally weary horse, said that his nag was too tired and too old to run away. He was mistaken.

After only the first year, new representatives in Congress who had not been party to the lengthy procedures to found a National Zoo started the debate anew. The budget was reduced, authorization to purchase

In the last two decades of the nineteenth century, American bison crashed from tens of millions to fewer than one thousand, almost all of them shot to death in the expansion to the west. Conservation efforts of the New York Zoological Society and the Smithsonian Institution helped to save the species from extinction. These schoolchildren inspect the first bison at the National Zoo, Washington, D.C., in 1899. Today, more than 100 years later, school groups continue to flock to the zoo. (Reproduced courtesy of the Smithsonian National Zoological Park. Photo: Francis Johnston.)

animals was withdrawn, and discussions opened to consider abolishing the new zoo.

The zoo was now in such dire straits that when a kangaroo happened to be for sale for seventy-five dollars, a local pet shop owner purchased it and traded the animal to the zoo for guinea pigs, at fifteen cents each. It took three years of breeding guinea pigs to clear the debt before the kangaroo belonged to the nation. One winter, the zoo displayed all the animals from Forepaugh's circus, with an agreement that they could retain any offspring born during the season. Two kangaroos and a lion were thus added to the collection.

Early buildings were as primitive as the zoo's acquisition standards. A tar-paper shed was built for the Forepaugh elephants, and there was only one general building, constructed in the cleared center of the site, for all the animals that required heat: big carnivores and their timid prey all together. The shed was still in use as an elephant house more than fifty years later, as were many other "temporary" structures. Not until the late 1920s did Congress appropriate funding for large and substantial structures. The first was $157,000 for a bird house, followed by $220,000 for "an Exhibition Building for reptiles, batrachians, insects, and miscellaneous vertebrates," known as the snake house. Architect Albert Harris, hired to design these buildings, immediately left for Europe to study zoos. He drew up the plans for the new buildings on the ship home.

HORNADAY MOVES ON

William Hornaday, who had set the idea of a national zoo in motion following his bison expedition, had been appointed in 1889 as the zoo's first designer and superintendent, but quit in understandable frustration after one year of trying to initiate his creative efforts with no political support. Devoid of a job, he did, however, have an important reputation in exhibit design and, moreover, in a particularly American aspect of it. A brief return to Philadelphia is required to examine that.

One of America's earliest natural history museums had opened in 1785 at the home of Charles Willson Peale, in Philadelphia, with a collection that ranged from zoology (including Benjamin Franklin's stuffed Angora cat) to botany, mineralogy to anthropology, and paleontology to the latest technology, such as the Eidophusikon, which showed moving images. Early museums such as this, which had first appeared in Europe in the sixteenth and seventeenth centuries, were known as "cabinets of wonders" (Weschler, 1995), where all manner of natural wonders were exhibited alongside works of art and ingenious human-made devices. This concept fragmented in the nineteenth century as the modern age began to emerge, breaking up into separate institutions that saw only separate facets of the world: natural history museums, zoological parks, aquariums, botanical gardens, and science and technology centers. Peale's idea of a diverse focus on animals and plants, both living and preserved, as well as stories and objects from different lands and peoples, passed with his death, and his collection was broken up. His concept would not emerge again until Bill Carr founded the Arizona-Sonora Desert Museum, in 1952, with its regional focus on the botany, geology, archaeology, anthropology, and zoology of the Sonora desert.

One of Peale's precedents did continue, however. Trained as an artist, he had a flair for creating attractive, even exciting exhibits. An enthusiastic visitor from Europe exclaimed that the museum was worth making the trip across the Atlantic. One of Peale's techniques was to paint landscapes at the back of glass-fronted display cages, showing the natural habitat of the animals presented. It was taken into a much higher dimension of drama when naturalist Carl Akeley, about a hundred years after Peale, began to create dioramas showing animals in settings of their natural habitat that were wonderfully realistic and perhaps the most dramatic manifestation of a museum presenting ideas rather than objects. It broke the slavish preoccupation of the collector with his jars and packets of specimens and introduced the notion of an artist presenting a concept. Akeley's dioramas for the Field Museum, Chicago, and the American Museum of Natural History, New York, contained many objects that were artificial, but the magic of his talent was based on a sound understanding of the animals and their homes, gathered from long periods in the wilderness. His muskrat exhibit for the Milwaukee Museum, which still exists, was the first true natural-habitat exhibit. Ecology in museum exhibition had begun, and it was a technique in which Hornaday had specialized. When he was hired by the newly established New York Zoological Society to become director and design coordinator for their new zoo, it was a perfect marriage of a person and an institution that shared the same high standards and broad conservation goals. Those standards and goals have become imbued within the society and remain healthy and vital today.

When a commission to select and locate lands for public parks submitted its report to the New York legislature in 1884, one of its conclusions was that "a park system that failed to include a zoological park would be wanting in one of its essential requisites." Some were of the opinion that the existing Central Park Menagerie fulfilled those requisites, but others were able to perceive the need for something more useful and inspiring. It took years of frustrations and prolonged and vigorous pressure from several influential people before the legislature approved an act to incorporate the New York Zoological Society in 1895.

Resistance occurred from some surprising places. Some politicians feared that a grand new zoo would lead to closure of the Central Park Menagerie, which was enormously popular with residents of the ghettos on the lower East Side, in the interests of property owners on Fifth Avenue. Others resistant to the idea were pet shop owners who believed the new zoological society intended to breed and sell small animals and put the pet shops out of business, and there were also rumors that the whole effort was just part of an attempt to move the Central Park Menagerie animals to some other place under the control of William Conklin, who had recently been fired as superintendent of the menagerie. There was truth in only the first of these concerns. During a fund-raising campaign in 1897 it was suggested that the Zoological Society might find favorable assistance from wealthy residents of upper Fifth Avenue "whose property would be greatly benefited by the . . . absorption of the Central Park Menagerie" into the new zoo. (The Zoological Society did take over the menagerie, but not until 1980.) The other fears were unfounded, based on no more than rumors and gossip.

The New York Zoological Society held its first meeting 7 May 1895, and the Bronx Zoo (or New York Zoological Park) opened four years later, in November 1899, following many protracted and difficult meetings of board members who argued interminably about many things, particularly a suitable location. But a committee headed by Columbia University professor Henry Fairfield Osborn, dean of American paleontology, and architect C. Grant La Farge had made it clear that they wanted the society to create an entirely new concept and a bigger and better zoo than had ever been attempted before.

The largest zoo in Europe was Berlin's 63-acre zoo, and the largest in America was the National Zoo with 166 acres. Osborn and La Farge proposed something in the region of three hundred acres. As William Bridges, in his history of the New York Zoological Society, accounts, they wanted a site where they could "place both native and foreign animals of the tropical, temperate and colder regions as far as possible in natural surroundings. Thus the larger wild animals of North America . . . should be shown not in paddocks but in the free range of large enclosures, in which the forests, rocks, and natural features of the landscape will give the people an impression of the life, habits and native surroundings of these different types."

The debate about a location was tabled until a director "of practical experience and acknowledged scientific standing" could be found. Meanwhile, William Hornaday had been so devastated by his frustrating year as director of the new National Zoo that he believed nothing could ever revive his interest in zoological gardens. The invitation that was to come from Henry Osborn, in 1896, to be director of the new zoological park in New York tempted him to reconsider: "The magnificent possibilities of your plan are enough to awaken [my] keen interest. The fact that your zoological park will undoubtedly be larger and finer than any other . . . quite stirs the blood."

Hornaday's first task as director was to select a site. Three weeks walking Van Cortlandt, Pelham Bay, Crotona, and Bronx Parks convinced him that the southern section of Bronx Park was the paradise he needed for "creation of a truly great and monumental zoological park." After settling this problem, William Hornaday and his wife set sail for Europe, at the request of the board, to investigate "methods of management . . . means of support, details and plans of buildings . . . special methods of caging and exhibition . . . photographs, plans, maps, architect's details, etc." They visited fifteen zoos, returning with $88.93 from their $500 travel allowance and tremendously enthused about the popularity of Europe's zoos.

Before the end of his first year, Hornaday produced a preliminary plan for the 264 acres of the Bronx Park site, and architects Heins & La Farge set about fleshing it out with buildings and infrastructure, detailing buildings that would be erected as soon as funds were available. The society, despite setting out to create a new type of zoo, made the usual error of copying from other zoos. Hornaday had brought back plans and photos, as requested, and the architects then modeled the lion and reptile houses after London's, the elephant house after Antwerp's, and the antelope house after Frankfurt's.

BOARD SKIRMISHES

Equally unfortunate, and sadly not unusual, was the precariousness of the authority that the board gave to their director in carrying out his responsibilities. Executive committee members, having filled their major responsibility by hiring a competent person as director, now persisted in reversing his decisions and directing him on matters of their own opinions. During construction of the new zoo buildings, Osborn sent instructions that the wires of the bird-house cages should be "a light color with perhaps an admixture of blue." The columns between the cages were to be "ivory white with an eggshell finish" and the "water closets should receive an inconspicuous green color." Present-day staff at some major zoos, perhaps most notably at San Diego, will recognize this sort of meddling all too well.

On another occasion, Hornaday is told not to "cut any bushes of any kind. Nor move any stone." In May 1899, when frantic building operations were underway and the first shipments of animals were beginning to arrive at the park, Osborn wrote, "I am in very bad humor about the border planting [and] much disappointed in the Prairie Dog enclosure." Hornaday replied, "When the fence is in position and the walk is built and the fence is painted the color it is to have, the entire effect will be totally different from anything you can get now. Visitors will look over the fence and not through it." On one occasion he came close to offering his resignation, after a board member had fired an assistant curator for disagreeing about the care and feeding of some baby animals (Bridges, 1974). Such examples of discord sadly persist in a few institutions today. Unfortunately, some zoo directors face the daily problem of managing a complex institution for which they are given full responsibility but little authority. Such situations complicate running some of the world's major zoos, sometimes to such an extent that it affects their quality and progress. Attempts by some zoo directors to try new directions or to pursue difficult agendas are often thwarted. There is abundant evidence, too, that at institutions where board members adopt a quasi-executive role, the staff invariably become fixated on *process* rather than product, retreating into activities that only pretend to justify their worth and existence. It is more than an irksome problem, and it lies close to the center of what prevents the advancement of some important institutions.

William Hornaday, first director of the New York Zoological Park, at his desk in 1910, from where he wrote his influential book *Our Vanishing Wildlife,* as well as his less effective tirades against those who persisted in calling the park the Bronx Zoo. (© Wildlife Conservation Society, headquartered at the Bronx Zoo.)

There is a potential remedy. Good systems of accreditation in some countries ensure that zoo management meets the standards of their respective national associations. If a similar system of accreditation were introduced for boards, it would assist board members in carrying out their proper (and very real) responsibilities and help check the perpetual—and very understandable—temptation to micromanage and make operational decisions.

The frailty of egos that can exacerbate problems between the executive and the board was nicely displayed in a letter to Hornaday from the board president, regarding his displeasure with an article in the *New York Post* that honored curator Raymond Ditmars for something new at the zoo. "We cannot hold our Members together . . . unless the members of the Society feel that they are getting the credit." But Hornaday, his patience repeatedly tested, had great resilience and stayed at the job for twenty more years. It was fortunate that he did. He became an ever more influential voice in wildlife conservation and steered the Bronx Zoo from its first days to its position of holding the highest standards.

THE BRONX ZOO

The New York Zoological Park opened 8 November 1899 with 1,157 animals on display within its five and a half miles of perimeter fencing. New York's wealthiest and most influential were there in force, and it did not take long for membership in the Zoological Society to become a prerequisite for membership in society. In spite of the Zoological Society, however, the New York Zoological Park has generally always been referred to by its nickname, Bronx Zoo. Hornaday waged a losing battle by continually attempting to change the situation. He informed the newspapers that "the nickname 'Bronx Zoo' is undignified, offensive, injurious, totally unnecessary, [and] inexcusable." He was unable to forestall the usage that still continues, even after the recent name change to Wildlife Conservation Park.

Hornaday fought more successfully for issues of wildlife protection. His book *Our Vanishing Wildlife* greatly influenced legislators to approve bills to protect migratory birds, for example, and he spearheaded many similar efforts to introduce wildlife conservation legislation. The New York Zoological Society paid for ten thousand copies of the book to be distributed to legislators around the country. Heavily engaged in a campaign against the fashion for decorating women's hats with stuffed birds and wild bird feathers, the society in 1912 purchased sixteen hundred hummingbird skins at a quarterly London auction for two cents each, glued them to cards, and mailed them to legislators and prominent women. The society's ingenious efforts were successful and new legislation was introduced that hastened the fashion's fall from vogue.

HORNADAY VERSUS HAGENBECK

With long years of experience as director of the New York Zoological Park, Hornaday was constantly asked for advice from new zoos across the country.

Usually he replied freely and copiously. Responding to an inquiry from the director of St. Louis Zoo about the merits of moated enclosures for bears, he said that in 1899 Hagenbeck

> laid before me the idea, original with him, of the so-called "barless bear dens." I pointed out to Mr. Hagenbeck that we intended to develop along scientific lines an educational institution in which . . . people might study to the best advantage the most important wild species of the world. I pointed out to him the great disadvantage that would be entailed by having our bears separated from our visitors by a distance of sixty or seventy feet. We deliberately decided against the Hagenbeck idea [and] we have never regretted it. I think the St. Louis Zoological Society is making a great mistake in putting all its money into costly piles of rocks and concrete to shelter far distant animals.

Hornaday's dislike of Hagenbeck's panoramas seems contradictory with his early enthusiasm for and expertise in developing habitat-based diorama displays and may have been fueled by anti-German sentiments he developed during World War I. In 1912 Hornaday had written to Hagenbeck, telling him that he had converted his office into a Hall of Fame and inscribed on it the name of "the greatest zoological garden builder . . . HAGENBECCK." During World War I, however, Hornaday came to hate the "German menace," and after peace broke out he wrote to colleagues across the country urging them not to buy animals from German dealers.

After thirty years of service Hornaday resigned, at the age of seventy-two, and spent his remaining eleven years engaged in conservation battles. One of his last acts, in 1937, was to send a letter to President Franklin Roosevelt asking him to use his influence to save the remnants of wildlife in the United States.

Reid Blair, a veterinarian and the zoo's second director, had worked under Hornaday for more than twenty-four years. Despite or perhaps because of that, he instituted many changes almost as soon as he took office. He immediately proposed the idea of "an experiment with the Hagenbeck idea of barless enclosures for animals," and received an intense and lengthy letter of chastisement from Hornaday. At the time, Emerson Brown, director of Philadelphia Zoo, was, with Hornaday's active support, leading a campaign against "Hagenbeckization" of America's zoos. Hornaday complained that "the Hagenbeck fad has inoculated some half-baked western zoo-makers."

Blair had no doubt learned to keep calm in the face of Hornaday's passionate temperament. He responded that Brown and Hornaday were "unduly alarmed about Hagenbeck's scheme to modernize American zoos. If we do not use up all our energy in trying to maintain that we have spoken the last word in zoological building and exhibition we may be able to pick up a few ideas worth considering from time to time." It was the first intimation of a recognition that the Bronx Zoo was becoming outmoded in terms of its exhibitry. But the stock market collapse of 1929 crushed any hopes of new works at the zoo. Blair's tenure there, from 1926 to 1940, could only be a hold-together era in regard to building and exhibit innovation.

NEW DIRECTIONS

Blair did institute one fundamental change, with an idea well ahead of his time. He wrote to the board: "We have stressed the entertainment and recreational features of the park . . . and should now develop the educational and scientific fields." He received permission to hire Claude Leister, a biology instructor at Cornell University, as the first curator of education. Establishing such a position in a zoo was remarkably farsighted.

The period of quiescence at the zoo between the Depression and the start of World War II changed with the arrival of Henry Osborn's son, Fairfield, as president in 1940. The younger Osborn had a vital, driving personality and held the post for twenty-eight years, leading the society to new heights of accomplishment in conservation and new directions for the zoological park.

Attendance had been declining at the zoo for several years. Many modernization plans had been proposed, but no construction funds were available. Although unable to change the zoo structurally, the new president initiated a new and more open attitude and kindled a friendlier atmosphere. He warmly encouraged initiative and experimentation. The zoo began to open itself up to new ideas. The forty-year ban on cameras in the zoo was abandoned. The uneconomical and badly run restaurant was closed, and a road train for visitors was introduced. A Department of Insects was formed and a series of art shows initiated. Steadily and quickly the place not only became more customer oriented and more accessible, but its financial resources began to increase as well.

One of the most important changes was a remarkable new exhibit, the African Plains, for which the new president secured anonymous funding from a benefactor later known to be department store owner Marshall Field. Opened in 1941, this exhibit was the catalyst for a sudden run of record-breaking attendance. More notably, in terms of zoo history, Hagenbeck's vision had been improved upon, lifting the concept to higher levels of realism. It may have been the first occasion that vegetation was manipulated to simulate exotic species: elm trees, mountain holly, and Texas water locusts were cropped to resemble acacias and other savanna species (Mitman, 1996). Significantly, however, and as with Hagenbeck's romantic panoramas, the human visitors continued to be placed outside the scene. They were still separated from the naturalistic landscape, looking out and into a stage set. If the time had been right, though, and if society had been ready to embrace it, this exhibit could have launched the zoo design revolution that had to wait until the public conservation awareness and the landscape-immersion exhibits of the 1970s. Nonetheless, it did precipitate a revolution on its home ground, setting the stage for a remarkable run during the past thirty years of new exhibits at the Bronx Zoo that have consistently set the highest standards.

Fairfield Osborn also reoriented the society to focus again on its old obligation of commitment to conservation, and he expanded it to include natural resource studies of forests, soils, and water. "If these primary natural resources continue to disappear," he explained, "the gains that have been made in wildlife conservation will be forfeited" (Osborn, 1941). His idea for a twelve-acre conservation exhibit in the zoo unfortunately never materialized, but he did succeed in establishing in 1948 a fifteen-hundred-acre conservation park at Jackson Hole, Wyoming. Designed to show in natural conditions the animals that the society was working to conserve in North America, it also provided a site for biological studies of the Rocky Mountain biome (the site was later absorbed into the Grand Tetons National Park). Also in 1948, the society established the Conservation Foundation, "to promote conservation of the Earth's life-supporting resources . . . for the sustenance and enrichment of life." Nicknamed "ConFound" by the staff, it was enormously productive, producing and distributing books, films, and pamphlets, and carrying out special projects from Jamaica to Alaska.

Large new exhibits were slower in arriving, hampered by labor and materials shortages. At last, however, in 1950, a great ape house was opened: the first major structure at the zoo since the zebra house in 1914. It was to be auspicious and ushered in a new era of exhibit development that has continued unabated, but it also presaged a sad event.

The exhibit contained outdoor yards for gorillas, chimps, and orangutans, with water-filled moats to separate animals and visitors. It was believed these would be safe and effective barriers because apes cannot swim. An accidental fall into the water, it was thought, was surely unlikely with such sure-footed animals, but as a safety measure, cables were fixed to the moat wall, strung just beneath the water line. On 13 May 1951, before a huge crowd, a male gorilla slipped while running and fell into six feet of water and almost immediately drowned. He made no attempt to use the cables or in any way try to save himself (Bridges, 1974).

Modifications were made, regrading the moats to create a gradual slope from a depth of only a few inches. It was one case where zoos would have done well to copy one another, but the lesson learned at the New York Zoological Park did not gain a response by others as quickly as it should have. Several more gorillas drowned in deep water moats in new exhibits at other zoos before it became obvious that they were death traps.

CONWAY TAKES THE LEAD

Today the Bronx Zoo/Wildlife Conservation Park contains more examples of progressive zoo exhibit design than any other, almost all of them based on concepts by William Conway, who joined the society as its Associate Curator of Birds in 1956 and retired as president in 1999. The World of Darkness, opened in 1969, and the World of Birds, in 1972, each designed by Morris Ketchum, were the first in this continuing renaissance. Ada Louise Huxtable, architecture critic for the *New York Times,* wrote that the Bronx Zoo "entertains, instructs and proselytizes, and it uses the tool of architecture to do so with singular skill. The World of Birds . . . is surefire drama and painless education. And fun."

To the everlasting glory of the Bronx Zoo, in the 1960s Conway set up an in-house design team. Some zoos, notably in Europe, such as London Zoo and Copenhagen Zoo, have employed staff architects for many decades. The difference at the Bronx was that now the designers were a multidisciplinary team, responsible for the exhibits, displays, and interpretation, not just the enclosing building. Notably, the team was originally headed by Jerry Johnson, a former theater set designer who brought a new emphasis to the concept of zoo design as theater. Johnson created the illusion of wild habitats by contriving zoo exhibits as stage sets, like gigantic dioramas, beautifully convincing and almost mystical in their visual appeal. His talents were apparent inside the World of Birds and the aquatic birds hall.

This tradition of zoo design as theater, as a stage for natural drama, has continued under the leadership of landscape architect (and ornithologist) John Gwynne, who since 1981 has overseen creation of many excellent new exhibits, including difficult refurbishments of heritage-listed buildings such as the old elephant house and others of similar vintage in Astor Court at the heart of the zoo. It is, however, the completely new exhibits developed by Gwynne and his multitalented design team, incorporating the skills of sensitive and imaginative designers such as Walter Deichmann, Lee Ehmke, and John Sutton, that are most exciting: particularly Jungle World, opened in 1985, the Himalayan Highlands exhibit, two years later, and the Baboon Reserve, which opened in 1990.

After many years of planning, the Bronx Zoo design team, in collaboration with David Helpern Architects, in 1999 completed a $35 million six-and-a-half acre Congo Forest exhibit, featuring two troops of gorillas in naturalistic habitat exhibits, similar to but larger than those first built at Woodland Park Zoo in the mid-1970s. This exhibition habitat, however, aims to go further. It displays not just gorillas but also colobus monkeys, Congo peacocks, okapi, mandrills, guenons, bush pigs, pythons, fishes, and several other small vertebrates and invertebrates, in extraordinarily realistic environments. The degree of complexity and beauty in this exhibit is rarely found in zoos. Its greatest difference, which any zoo could emulate, is the extent to which it is committed to encouraging visitors to directly participate in efforts to save wild habitats in central African forests. In recent years the Bronx Zoo has been urging all zoos to move away from being mere exhibitors of wildlife and instead to become conservation centers. William Conway has produced lists of ideas that could help them reach this goal (1995c). He has experimented himself with one of the ideas at the Congo Forest exhibit.

At the entrance to this exhibit, visitors are required to pay an admission fee. As they walk through the exhibit they see contented and relaxed animals in lush and natural-looking habitats, but they also will see examples of human-caused environmental destruction. From one end of the exhibit experience to the other, visitors are challenged, informed, persuaded, exalted, saddened, and delighted. At the conclusion, the visitors are asked to vote how their admission fees should be used by choosing from a selection of conservation actions. The monies are then so directed. It is an inspired (and inspiring) example of encouraging public involvement, commitment, and empowerment. In addition, the Wildlife Conservation Society has embarked upon a $20 million campaign to apply directly to saving wild habitats in the Congo.

The Baboon Reserve at the Bronx Zoo, two acres of simulated high-altitude Ethiopian grasslands adjacent to an ersatz African village built in the Somba style of architecture, displays gelada baboons, Nubian ibex, rock hyrax, and blue-winged geese in a delightfully realistic landscape, which is as skillfully interpreted as it is designed. Two groups of female geladas and their babies, each congregated around a single adult male, duplicate the multileveled organization of this species' natural sociology. Visitors can thus watch group dynamics and patterns of foraging as they would occur in the wild.

The obvious stars of the Himalayan Highlands exhibit are the snow leopards. They were the last residents of the old barred cages at the Bronx, and now leap around, at least on brisk wintry days, in a simulated mountain forest habitat. Or else they sleep for hours, digesting their heavy protein meals, lounging unashamedly on heated rocks placed close to viewing areas, separated from human visitors only by piano-wire screens. People approach these viewing areas along a narrow rocky path, under a massive boulder, through bamboo thickets, and past vernacular architectural structures such as stone bridges, and then find the animals in what appears to be an

unconfined area, contained only by a gauzy aviary-like structure that is virtually invisible among the trees.

There are many satisfying aspects to these Bronx Zoo exhibits: the attention to detail, the striving for authenticity, the consideration of animal needs, and most importantly the fact that all these exhibits are driven by and formulated around aspects of conservation. Perhaps most wondrous, though, is the fact that people in one of the most densely populated urban environ-

A snow leopard and a visitor peruse each other at the lion house in the Bronx Zoo, one chilly February day in 1906 (facing page). L. P. Hartley (1958) said that the past is like a foreign country: "They do things differently there." But things remained the same at this and at other lion houses in zoos around the world until recent times. In some zoos, equally barren and even worse conditions still prevail. Baby snow leopards at the Bronx Zoo in 1986 (above), however, thrive in a naturalistic exhibit designed by zoo staff and landscape architect John Gwynne. (Photos © Wildlife Conservation Society, headquartered at the Bronx Zoo.)

ments on the planet are able to so easily immerse themselves in such convincing replicas of wilderness habitats. New York contains and displays many of the world's finest cultural treasures and represents many of the greatest human achievements in the modern age. It is, then, all the more impressive that New York also has the capacity to remind its citizens and visitors of the wonders and glories of the natural world and to reveal these phenomena so convincingly.

The consistent quality of the exhibits, the depth and range of the education programs, and the vitality of the conservation projects emanating from the Bronx Zoo are very much the product of its recently retired director, William Conway. In his first annual report in 1961, Conway expounded a philosophy that remains so sound every zoo director should read it regularly:

> Zoo visitors should have the opportunity to learn something about each animal's environment through natural habitat displays, to explore the mysteries of wild animal behavior, [and] to be informed by special displays. . . .
>
> The justification for removing an animal from the wild for exhibition must be judged by the value of that exhibition in terms of human education and appreciation, and the suitability and effectiveness of the exhibition in terms of each wild creature's contentment and continued welfare. Man's works and his proliferating population impose new stresses upon the wild species, and those stresses demand greater responsibilities from other zoological parks of the world.

CHAPTER SIX

IMMERSED IN THE LANDSCAPE

The 1960s saw people in the Western world, North Americans in particular, reassessing their lifestyles and experimenting with new ideas. The natural environment became a focus of concern, especially the fate of wild places. Also, America was approaching the end of an era during which an accepted pattern of living and of shaping the land that had been transferred intact from northern Europe was being seen by some as increasingly irrelevant. The United States had inherited an English philosophy of taming the land, but it was an ethic clearly unsuitable for the American West. That ecosystem cannot be plowed and grazed by large herds of domestic livestock without suffering a high degree of ecological degradation. It is land that needs to be approached with an intent to cooperate, not to subdue. The need for this type of approach is one of the reasons that American Indians had a different attitude toward the land than the European-heritage American ranchers.

When ecology emerged as a matter of public interest about a quarter of a century ago, a few zoos began to consider making conservation their central role, with Gerald Durrell of the Jersey Zoo, George Rabb of Brookfield Zoo, and William Conway of the Bronx Zoo leading the discussion. It was becoming clearer to more people that zoos as mere entertainment facilities were not justified. Conservation, and education about conservation, was to be paramount. A few zoo professionals, especially those associated with the growth and development of the Bronx Zoo, had pursued this role much earlier, but at this point the philosophy became more widely accepted and developed quickly. The general media routinely began to carry articles explaining the threats to wildlife that had risen from habitat eradication. Conservation and environmental activist organizations were increasing membership and building political clout. Through the 1970s, zoo professionals became increasingly aware of the need to engage themselves in conservation programs and the American Zoo Association asserted that conservation had become its highest priority.

In addition, with the stringent philosophies of the modern movement in architecture beginning to degrade, and the public sentiment for conservation and concern for wildlife beginning to swell, the stage was also set for bold improvements in zoo design. There was growing public concern about zoo conditions in many countries, although responses in each reflected the strangely varying attitudes toward wild animals that are found among different human cultures. In the USSR, the Marxist notion that reverence for Nature was a capitalist aberration had long held sway, and Stalin had promoted wilderness as an enemy of the working class; wild lands must be tamed and made to serve or be eradicated (Weiner, 1988). Thus, the dedicated few who expressed uneasiness about the wretched conditions of their zoos ran into the impenetrable walls of a cynical bureaucracy and retreated in frustration. In northern and western Europe, public disquiet tended to translate itself into either lethargic negativism from intellectuals who shunned zoos as places to be discarded after childhood, or it sparked organized activism from those with sensitivity to animals who saw zoos only as prisons. Neither strategy was successful to improve zoo conditions and did no more than cause anxiety among European zoo directors. In southern Europe and the Latin countries some of the worst zoos in the world continued their atrocious traditions with no comment or disturbance from the populace. The problem of the generally mediocre-to-awful zoos that proliferated in Japan never even appeared on the menu of public issues. But in the United States public demand for a new emphasis on the quality of life for animals in zoos grew steadily more insistent, and with increasing volume. It was a propitious time to instigate new, even radical, ideas, and Seattle proved to be a perfect place for the attempt. The city had a history of progressive reformism, a pleasantly mild climate, an approved bond issue for a new zoo master plan, and a seventy-year-old zoo on a ninety-acre wooded site that was desperate for attention.

Most zoo plans of this era were based on taxonomic arrangements, or, occasionally, had animal exhibits arranged according to geographical origin. Only rarely were exhibits habitat based: the aquatic birds hall at the Bronx Zoo, the Life Underground display at the Arizona-Sonora Desert Museum, and the Wolf Woods at Brookfield Zoo were the only significant examples.

IMMERSION IN SEATTLE

The 1976 long-range plan for Seattle's Woodland Park Zoo may have been the first such document to declare that the animals were the primary client, and their needs took greater priority than those of the keepers or the visitors. Jones & Jones, the landscape architects employed to prepare the plan, responded enthusiastically to such a challenge, for the basic idea was not unknown to them. Grant Jones had already developed a design ethic that Earth was the primary concern to be addressed by landscape architects and that the firm's designs must place Nature first.

Described by *Landscape Architecture* magazine as "trailblazers of the highest order" (Powell, 1997b), the then-new firm of Jones & Jones had not designed any zoo exhibits but, building on the revolutionary studies of landscape architect Ian McHarg, they had developed a groundbreaking methodology for reading a landscape to determine natural process and form. The firm's first major project had been to develop a comprehensive evaluation of the dynamics of the Nooksack River, which included an inventory of landforms that identified no fewer than fifty-nine units of stream form and topography, the characteristics of each plotted against a matrix of natural, cultural, and aesthetic values. The results were new standards for landscape analysis and a perfect approach for Seattle's new zoo study.

Consciously departing from traditional taxonomic and geographic approaches, Woodland Park asked Jones & Jones to test the feasibility of reorganizing the zoo on the basis of bioclimatic zones. Two years earlier, in 1972, ecologist Leslie Holdridge had devised an elegant system for classifying the world's habitats in such zones, based on temperature, precipitation, and evapotranspiration—the interacting variables that create the zonal patterns of the living membrane, powered by the sun, that cloaks planet Earth. The varying confluences of these parameters confine plants and animals to their specific parts of the world. By constructing a triangle using these three parameters, any place on Earth can be located within the Holdridge system of life zones.

Jones & Jones researched almost every conceivable aspect of the Woodland Park Zoo site, diagramming shade and sun patterns, slopes, natural drainage, soil types, vegetation cover, and notable trees and extrapolating other considerations from these data to test against the Holdridge system. They were then able to select ten bioclimatic zones for potential replication. Unlike all previous zoo plans, the resultant layout responded directly to the natural microhabitats of the site and objectively reflected the specialized habitats to which plants and animals had adapted over the passage of time.

In habitat-based zoo plans, as opposed to taxonomic layouts, ecological areas are subdivided by geography. A section devoted to grasslands, for example, would not sensibly display rhinos, pronghorns, and kangaroos together, in spite of their all being grassland species, because they are from Africa, North America, and Australia, respectively. It is useful, however, to also remember that zoogeography offers a merely transitory view. Marsupials may now be seen as characteristically Australian, and rhinoceroses as characteristically African (although they arrived later there than anywhere else). But marsupials and rhinos originated in North America, about 120 million years ago. Elephants once ranged the entire globe, apart from Australia, including (in the form of mammoths) North America, until just a few thousand years ago. Modern elephants originated in Asia and migrated to Africa. Lions once ranged throughout Eurasia and deep into China. By definition of origin, neither elephants nor lions are African animals. Wild animal populations are not placed in or restricted to some particular geographic area. They live on a dynamic planet. Where we discover them is merely where they happen to be during our time on Earth. Colin Tudge (1995) speaks of "cascades of animals migrating, radiating, moving on or drifting back from whence they came."

Because many different types of vegetation can be grown in Seattle's relatively mild climate, several bioclimatic zones could be incorporated in the zoo's long-range plan. Other zoo sites might be restricted by less moderate climates, a fact that could nonetheless prove to be advantageous for them, by leading to specialization. At the least it should help ensure that the zoo works within its own natural confines in terms of vegetation and the accuracy of the habitats it can simulate.

Woodland Park Zoo's plan (Jones, Coe, and Paulson, 1976) was based on vegetation patterns to a critical degree, not just on animal species, and this is evident in many details. For example, a forest-edge habitat was deemed essential to make a logical transition between Taiga and Tundra zones on the zoo site. Caribou and musk ox were designated as linking zoo exhibits within that transition. These animals, however, were not part of the zoo's collection and possibly never will be. Nonetheless, the plan called for planting that part of the zoo grounds as caribou and musk ox habitat, ensuring authenticity of continuity from one bioclimatic zone to another. Musk ox and caribou habitat has its own special fascination and its own lessons to impart, whether or not the animals are visible. For the first time, a zoo was stating that the presence of animals was not necessarily the primary object of the exhibit.

MATTERS OF MICE AND JAGUARS

The Arizona-Sonora Desert Museum reached the same conclusion on two occasions in the early 1990s. The first occurred when the museum was planning a desert grasslands exhibit, and consultants suggested including pronghorns. For various reasons, including difficulties with health care for these animals in captivity, staff advised against this. They wanted to concentrate on replicating the habitat rather than devoting limited resources of space and money to a pronghorn exhibit. Some trustees complained that not including the pronghorn would exclude from the exhibit "one of the most interesting animal species" of the region. Pronghorns are indeed the largest animals found in the desert grasslands, but size equates only with visibility, not interest. There is, for example, a curious little mouse that has the unmouselike habit of eating other diminutive animals. This small creature, called the grasshopper mouse, also stands on its hind legs and howls at the moon, in high-pitched squeals, like a tiny wolf. It is as good a candidate as

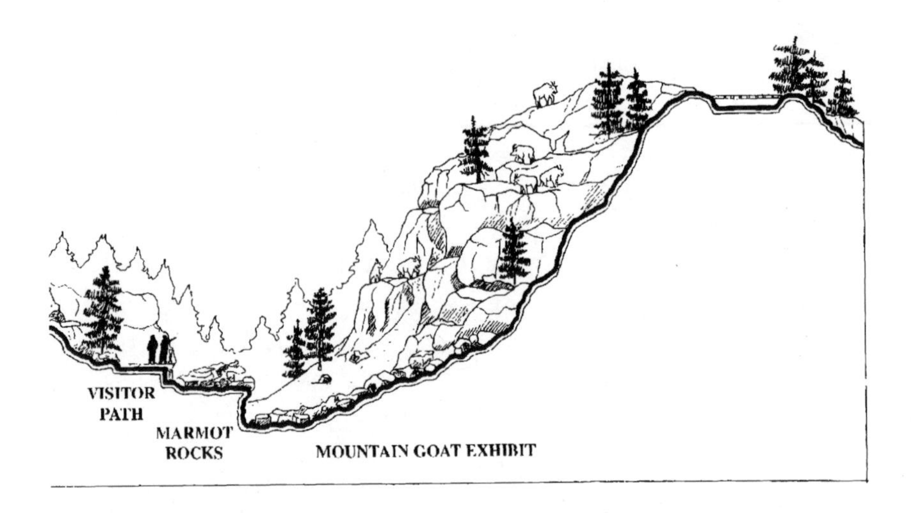

The grandness of scale in Jones & Jones's long-range plan for the Woodland Park Zoo exploited the existing characteristics of the site, and their proposed exhibit for mountain goats, if constructed, would have set a standard for zoos in excess of even Hagenbeck's vast panoramas. The mountain goat exhibition habitat called for a series of peaks and ridges with long sweeping talus slopes and abundant sparkling rivulets. (Drawing courtesy of Jones & Jones, P.S.C.)

a pronghorn for "most interesting animal" of the desert grasslands. More pertinent, studies reveal that certain mice are keystone species in this habitat: their feeding habits prevent it from reverting to scrub (Brown and Heske, 1990). The new exhibit therefore gives its greatest space and attention to a large number of desert grass species, attempting to illustrate the special drama and characteristics of this maligned habitat. It does this without the charismatic pronghorn, but with masses of equally photogenic grasses. And with a life-size bronze statue of the grasshopper mouse.

Later, staff at the Desert Museum were debating whether or not to replace a deceased jaguar who had died of old age. Museum management did not wish to acquire another jaguar, believing it could not satisfactorily meet the needs of the species and could in any case carry more effective ecological stories with very small species that were better suited to captivity and less expensive to house and care for. Several board members, urged by some past donors, strongly disagreed. One said the jaguar was the "most important, most magnificent" animal in the Sonora Desert biome. Ecologically, however, many other animals in the desert are more important than the jaguar.

Staff responded with a suggestion that there was another way to solve the problem. The Sonora Desert grades into tropical deciduous forest at its southern edge. This is one of the most fragile and threatened habitats on the planet. Staff proposed that a tropical deciduous forest exhibition habitat be created at the museum, the entry into which might lead visitors to expect to encounter a jaguar. Signs, for example, would explain that this habitat was home of the jaguar. Evidence of a jaguar's presence would be present: footprints, remains of a recent kill, maybe an occasional grunt or distant jaguar cry, eyeshine in the back of a dark cave. An inconspicuous side trail would lure away the inquisitive visitors who might notice it and lead them to an opening in a cave from where they could see visitors along the main path on the other side of steel bars. The intent was to give the adventurers the impression that they had somehow penetrated a jaguar enclosure.

At this point a sign would explain that there was no live jaguar in this exhibit, because it is not the most important component of the biological community. To arouse attention, if not ire, the sign would point out that human interest in such glamorous species is almost entirely aesthetic and emotional and is not based on the significance of their ecological roles. The sign would then go on to say: "Now that we have your attention, the rest of this exhibit will demonstrate why the tropical deciduous forest is an important and fascinating habitat, regardless of jaguars." It would have been an opportunity to initiate a debate about the need for zoos to concentrate on habitat and ecosystem conservation, not just preservation of large and glamorous species. No doubt many people would have missed the point of this risky exhibit and simply felt cheated out of seeing a jaguar. How many visitors, though, get any sort of message from all the live jaguar exhibits in zoos? Ultimately, however, this novel display was never constructed but neither was the dead jaguar replaced with a display of a live one. Its miserable old cage was remodeled as an interpretive shelter.

COMMITMENT AND AUTHENTICITY

The matter of visibility of animals raised many new problems for Woodland Park Zoo as it began to develop its new plan and attempt to construct another of its innovations—the concept of "landscape immersion," a term coined by Grant Jones. The zoo had asked Jones & Jones to simulate habitats in a landscape of more perfect illusion than had been achieved before, with no sense of separation between animals and people. It specifically requested that they blur the barriers by putting the same landforms and plantings in both the public and the animal areas. The intention was that by exhibiting animals in landscapes that closely resembled their natural habitat in every possible detail and by immersing the viewer within that same wild habitat, people would subconsciously make connections between the interdependence of certain animals, plants, and habitats. Landscape architect Lee McMaster, in his study of landscape design in American zoos, says the "colorful history of past zoo design paved the way for this new concept," an evolution from Hagenbeck's panoramas. Anne Elizabeth Powell, editor in chief of *Landscape Architecture* magazine recognized another aspect of the concept, explaining in her article "Breaking the Mold" that the landscape-immersion approach was "an astonishing departure from conventional zoo design because it reflected a pronounced shift in philosophy" from the homocentric to the biocentric view, and noted what Jon Coe, one of the planning team members, explains as "the basic conservation approach that says we just live here, we don't own the Earth, the Earth owns us." Coe added that the new approach allowed the viewer to become physically and psychologically immersed in the simulated created habitat. It appealed, he says, "first to the emotion and secondly to the intellect" (Coe, 1995).

The world's first landscape-immersion zoo exhibits were a gorilla exhibit and an African savanna at Woodland Park Zoo, both developed in the 1970s. There is now an ever growing and wider range of landscape-immersion exhibits in zoos around the world, notably at the Bronx, San Diego, North Carolina, Portland, Tacoma, New Orleans, Toledo, Miami, Cincinnati, Minnesota, Brookfield, Orlando, Zurich, Calgary, Melbourne, Perth, and most successfully, because of their natural surroundings and their commitment to absolute authenticity, at the Arizona-Sonora Desert Museum in Tucson. To be at all successful, landscape immersion must make a dedicated effort toward realism. Selected plants must be authentic or at least very closely simulate the appearance of vegetation types in the natural habitat. Replicated rockwork, if it is used at all, must faithfully mimic the form, colors, and textures of natural geological features that would be found within that habitat. As is too often demonstrated, the illusion of wilderness is easily shattered by discordant details.

The spectacled bear exhibit at Zurich Zoo, designed by landscape architect Walter Vetsch, opened in 1995. This is a rare example of landscape immersion in European zoological parks. Moreover, the degree of naturalism and the scale and drama of this exhibition habitat is of an extraordinarily high standard for any zoo. (© Zoo Zurich.)

Equally imperative, the claim to landscape immersion (the term already sadly becoming a hackneyed phrase in the zoo profession, just as zoos have degraded the word "habitat" to a substitute for "cage") must not forget its purposes. Even if visual authenticity is obtained, it must not be gained at the expense of the animals' needs. The original reasons for the concept required animal exhibit space to be as accurate a replication as possible of the natural habitat.

It seems that humans often fail to look beyond the superficial. A recent issue of *Landscape Architecture* featured the "revolution" of landscape immersion in zoo design, correctly identifying that it depends upon "immense complexities in habitat re-creation" (Powell, 1997a). It is therefore puzzling that the first two illustrations selected by the photo editor for this article are both views of the Lied Jungle, a glass-roofed structure that opened in 1994 at the Henry Doorly Zoo in Omaha.

The first photograph is a view looking down onto a tapir, apparently in a jungle clearing by a river edge, except that there is far more cleared space than jungle vegetation, and dirt and concrete appear more prominently than plants. If visitors could see behind the scenes here, they would realize how extremely cramped and sterile the holding spaces are for the animals, even by the typically inadequate off-exhibit standards that exist in most zoos. Even the exhibit spaces in the Lied Jungle, one of the world's largest zoo rain forest exhibits, are minute. Worse, these tiny spaces contain no natural elements. Almost everything the animals come into contact with in this "jungle" is made of concrete, steel, or epoxy plastic: unyielding, unchanging, rigid, and useless to the animals. There is nothing with which they can interact, nothing malleable, chewable, or moveable to serve as diversion, nor is there anything at all that is reflective of their natural environment. Even the animals' climate, which is controlled and sealed off from the outdoors, has no temporal qualities.

The second photograph in the magazine article showed a view of this same rain forest exhibit looking toward a public viewing space. The whole is so far removed from the mood or complexity of real rain forests one wonders why a landscape-design journal selected it as an example of anything other than a demonstration of what *not* to do. Fanciful artificial rock formations pile up to form the viewing platform from which visitors can survey

Tapirs huddled against a concrete wall (textured to look like rock) beneath a concrete tree (textured to look like wood) are marooned on a pitifully small slab of concrete (textured to look like mud) inside the $17 million Lied Jungle exhibit at the Henry Doorly Zoo, Omaha, opened in 1994. Visitors peer down upon the animals from high above. (Photo: Larry Vogelnest.)

a scene that destroys any sense of exploration or immersion and puts them at immediate eye level with an inelegant steel structure enclosing the fabricated whole. A waterfall surging inexplicably from the top of a perimeter wall reminds one of Vicki Croke's sadly accurate observation, in her study of American zoos, that most shopping malls build fountains more convincingly than zoos (Croke, 1997).

Henry Doorly Zoo's jungle is not significantly or essentially different from the other large and expensive rain forest exhibits springing up, particularly in many other American zoos in recent years, such as those in Denver, Portland, Pittsburgh, Baltimore, Fort Worth, Brookfield, Oklahoma City, Albuquerque, Boston, Tulsa, Washington, D.C., and Cleveland. Sadly, they all tend to look vaguely alike, planted more like a decorative conservatory in a botanical garden, with an obsession for bamboo, palms, and banana, rather than the untidy complexity of a rain forest. They are generally enclosed by heavily engineered reinforced concrete and steel structures with artificial rockwork as the animal "habitats." Serpentine paths lead visitors past exhibit spaces that, like the Lied Jungle's tapir exhibit, are usually devoid of anything useful to or natural for the animals.

We have long had accurate descriptions of tropical rain forests from observant travelers. Charles Darwin, like others who have experienced the forests, repeatedly spoke in his *Journal of Researches* of the "universal silence" of the deep forests of Brazil; in his 1996 study of Victorian scientific travelers, Peter Raby noted that Richard and John Lander, exploring West Africa in 1830, described the stillness of the forest, where "the song of birds was not heard, nor could any animals be seen," and that Henry Bates, living in the Amazon forest in the 1840s, spoke of its "noiseless solitude."

Modern zoo rain forest exhibits bear virtually no visual, ecological, or other resemblance to real tropical rain forests because they are not modeled on real ones. The modern exhibits are a perpetuation of the old zoo problem of enclosing too many animals in too little space and of merely copying each other's ideas. The designers and curators do not go to rain forests or other wildlife habitats and analyze these places' changing moods, distilling the myriad details that create a sense of wonder, spirit, and mystery. If they do make such visits, they seem to focus only on the animal species in which they are interested and return to their zoo with no full understanding of the sense of the *place.*

Jungle World at the Bronx Zoo, opened in 1985 and thus one of the first tropical rain forest exhibits, is by contrast an exhibit that accurately simulates the sense of a rain forest environment by using exhibit design as theater. Visitors move through a carefully orchestrated set of staged scenes amid the sounds, smells, and sensations of a steamy jungle atmosphere. This is consciously *not* an immersion experience; the visitor is clearly in a public space looking into an exhibition area. A real rain forest, with its incredible diversity of life forms, massive height, layering of vegetation, low light levels at the floor, and a forest canopy that is itself another continent of life cannot be reproduced as a full-size facsimile. The theatrical illusions of Jungle World therefore seem the intelligent alternative. Exquisite attention to detail is matched by the design team's skill in controlling sight lines. Unlike most conservatory rain forests, in which visitors physically and visually soon discover the confining limit of the building, there are cleverly controlled vistas in Jungle World that give the illusion that sometimes one is looking into a forest of limitless distance. It is put together with such competence and care that it suggests that it is a privilege to enter this rare and exotic environment. It is a triumph of design and a magical embodiment of an important philosophy. Tropical rain forests are disappearing at an alarming and frightening rate. Visitors to Jungle World, made aware of the complexity and delicacy of this ecosystem, can realize some of the consequences of such destruction and perhaps be motivated to help in rain forest conservation.

In theory, the superb dioramas of natural history museums should match the success of Jungle World. Carl Akeley's modeled habitats for African mammals at the American Museum of Natural History, for example, are more perfect in detail than a zoo could possibly achieve. The sight lines of the dioramas are perfectly controlled. Visitors can stand before one and almost believe that they are looking into a frozen moment of actual Africa. It fails empathetically, however, in that it impresses the eye and the intellect, but falls short of creating a sense of wonder for wild animals or a sense of urgency to conserve wild habitat. Is it because we know that the natural history museums' animals are dead? We are sadly too aware that the exacting precision of the taxidermist is married to the equally precise skill of those

who took such careful aim to avoid blasting the whole head off the baby warthog, who shot to death on separate mornings the lion and the zebra now forever locked together, and who squeezed the trigger with such delicate perfection that the bullet penetrated a part of the gazelle's fragile body where it will never show. Maybe, amid these dark halls of equal shame and wonder, it is impossible to connect with *life* as one can in Jungle World.

Natural history museum displays can be things of great beauty and fascination and can even transport one to distant times and places, but they cannot provide the same emotional connections as living plants and animals. As the World Zoo Conservation Strategy of 1993, written by the Captive Breeding Strategies Group of the International Union of Directors of

This nineteenth-century engraving of the Brazilian rain forest by Moritz Rugendas (facing page) was described by Charles Darwin as "most true and clever" (Raby, 1996). It accurately depicts the rain forest's rich complexity and diversity. Contrast this with the $6 million rain forest exhibit at the Fort Worth Zoo (above), opened in 1992. Described by the zoo as a "natural exhibit . . . in a lush tropical forest" (Ferguson, 1992), we can only guess the description that Darwin would have given. (Engraving courtesy of the Bodleian Library, University of Oxford; photo courtesy of the Fort Worth Zoo; photo: Barbara Love Logan.)

Zoological Gardens (now known as the World Zoo Association), points out, the fact that zoos exhibit living animals underscores the essential differences between them and other similar institutions, "and is what gives zoos their own unique character." It is also, of course, what gives them their own unique, even awesome, responsibilities.

The Lied Jungle exhibit cost almost $20 million. The Cleveland Zoo rain forest exhibit cost $30 million, as did the Bronx's Jungle World. These are not uncommon figures for rain forest exhibits. One of the first complete immersion exhibits, the African savanna exhibit at Woodland Park Zoo, hundreds of times bigger than any of these, cost just $4 million. In large part, this much lower figure resulted from its being an outdoor exhibition space not requiring the high cost of a building structure. The comparison therefore may seem ridiculous, but it contains an important moral. If zoos will focus more on exhibits that fit within the confines of their natural environment, dealing with habitat zones that lie close to their own place in the Holdridge system of life zones and rely upon landscape rather than architectural solutions, they can achieve much more for much less.

The African savanna exhibit is one of the bioclimatic zones in Woodland Park Zoo's long-range plan. Within each simulated zone, the zoo wanted to create naturalistic habitat displays of complete visual authenticity. This meant detailed specifications for terrain, soil type and color, and vegetation. The intended result was to make the exhibits as close in appearance to wilderness as possible. Initially this brought novel problems. Many visitors did not like the outcome, especially those accustomed to visiting zoos in which the lawns were neatly mowed, the shrubs tightly trimmed, and the animals in close proximity and in full view.

In the new exhibits at Woodland Park, wild grasses were planted and allowed to grow tall and shaggy, vegetation was fertilized and grew rampant, trees were planted in front of viewing areas to create the sense of looking through and into the habitats. Public spaces, once obsessively neat, now replicated wild places in mood and content. But complaints started to flow, first as mutterings of discontent resulting in signs being placed around the grounds explaining "These are not weeds!" Public dissatisfaction nonetheless persisted and zoo management began to hear reflections of concern from the mayor's office. Fortunately, as news began to spread that fundamental changes were underway, a new clientele started visiting the zoo. People who had avoided the place for years, depressed by its sterility and banality, began to return. Seattle newspaper publisher David Brewster described a trip to the zoo: "It was a glorious summer evening, so we went for a walk in the savanna; clear sky, warm breeze, late sun lying flat across the dry grass, and total silence, except for a few bird calls and the sound of our own feet on the gravel path. And then suddenly we heard a pounding of hooves, and over a little hill beside us came zebras, galloping for the pleasure of it, over the hill and down the other side and out of sight. I'll never forget that experience as long as I live" (Downey, 1980).

Soon a much larger number of visitors were expressing similar if less eloquent enthusiasm for the changes. The animals may have been more difficult to see, either farther away or obscured among dense vegetation, or even running out of sight, but when one did see them, the views were much more satisfying, the overall impression more rewarding and realistic. Attendance started to swell and the mayor's office began to receive more positive comments than negative ones.

IT'S A JUNGLE OUT THERE

The most significant change to the zoo, which was also to have effects upon zoos worldwide, was the development of a new gorilla exhibition habitat with live trees and lush vegetation. Designed by Johnpaul Jones, this exhibit, says Powell (1997b), "rewrote the definition of the term *zoo.*" Yet it also met strong initial resistance, both locally and within the zoo community, and many European zoo directors are still suspicious of the value of this design approach. This negativity was at first so strong and sustained that had it not been for the solid support of parks superintendent David Towne, who had earlier given the design team license to reach for new standards, Woodland Park might never have built the gorilla exhibit.

When plans for the exhibit were shown at an annual meeting of the American Association of Zoological Parks and Aquariums, the general reaction was hostile. No one had kept gorillas in exhibits with live vegetation, and it was suggested that attempts to do so now were ridiculous. Only one year earlier, the director of Houston Zoo, John Werler, had declared that "landscaping with live plants is impossible to maintain in a Great Ape enclosure where the animals are within reach of the vegetation; they very quickly destroy anything that grows" (Werler, 1975). One zoo director tried to prove the point by placing a potted palm in a cage then letting the gorilla in. He triumphantly explained to Woodland Park staff that the animal had immediately pulled the plant out of the container, tried to eat the entire thing, and then vomited. Many others as well, though without such weighty empirical evidence, prophesied that the gorillas would indeed destroy all the vegetation, turning the exhibit into a seasonal mud or dust bowl. Some also pointedly reminded the zoo that gorillas were highly endangered species and that it was irresponsible to allow them access to mature trees from which they would assuredly fall and crash. In addition, strong skepticism was expressed among many major zoo curators that landscaping the public areas with so much dense vegetation was a waste of money.

A major opponent was a local activist, Benalla Caminitti, employed at the University of Washington's Primate Research Center. She vehemently opposed the changes at the zoo but was particularly agitated about relocation of the gorillas into a naturalistic outdoor enclosure; she warned of dire problems with diseases from living in such an environment. The zoo's veterinarian, James Foster, did not support this opinion. He believed the gorillas would benefit greatly, both psychologically and physiologically, from living an outdoor life.

Caminitti was fortunately ignorant of a potentially serious problem that could have been fatal ammunition in her hands. Strictly speaking, the zoo did not have officially approved funding for the exhibit. A voter-approved bond issue ten years earlier had earmarked monies for specifically stated "zoo improvements." This list included a new primate house, but made no mention of gorillas. Woodland Park, like most other zoos, kept their gorillas in an ape house. The ancient primate house contained monkeys and lemurs. With no intention of building a new "house" for any primates and unwilling to accept the gorillas' continuing existence in their miserable quarters, the new zoo management team interpreted the dictates of the bond issue as liberally as possible, stretching the designation of "primate house" to mean "gorilla habitat" as well as exhibition habitats for monkeys and lemurs.

Even so, funds were extremely limited. About $500,000 was allocated to build the exhibit (when San Diego built their gorilla exhibit in the late 1980s, closely modeled on the Woodland Park exhibition habitat, they allocated $250,000 for the interpretive sound system alone), and to stretch those dollars the zoo decided to remodel existing bear dens and grottos to create the gorilla exhibition habitat. The dens were lit with new skylights and acoustically treated on floors and ceilings and at doorjambs to create a softer and gentler environment. Access doors were cut so that the gorillas could select where (and with whom) to sleep at night and choose to make nests of thick piles of hay in hammocks or on the floor.

Outside, the moats that had separated people and bears were filled in to greatly extend the animal area and thereby encompass some mature maple trees within the animals' range. The land was formed to create berms, so that the gorillas could get out of sight of each other, or of people, if they so wished, thus alleviating stress. Masses of plantings were installed, artificial boulders were built with heat pads beneath the surface to create warm sitting places on chilly days. The adjacent public areas were as large as the animal spaces and similarly covered with dense plantings. A narrow pathway of natural substrate snaked through this vegetation, bringing visitors to the display area almost as a surprise discovery in a forest clearing.

When everything was finished, it was all left alone for a year to allow the plantings to set down strong roots before being tested to possible destruction by the gorillas. It was a tense yet exciting time as the exhibit daily grew more lush and verdant. Most frustrating for zoo staff, the gorillas, locked in their blank box not far away, could not be informed that liberation was due.

In Woodland Park Zoo's 1976 long-range plan, each new exhibit is sketched out and technical details listed, such as a specification of 70 percent fine sand and 30 percent compacted sawdust as a soil mix for the gorilla exhibit. Of special value, however, is a word picture of each one. Ecologist Dennis Paulson had determined the physiognomy of each exhibit to match real, natural habitats, and Jon Coe, who had written his master's thesis at Harvard on animal environments, described them in words that sounded almost like poetic or romantic visions. The description for the gorilla exhibit includes notes on many aspects: here are those for "terrain" and "vegetation," as examples.

> TERRAIN
>
> [This will be a] steep-sided stream-cut valley with a few huge worn artificial granite boulders. A freshet will flow from an unseen source up a side canyon and be intermittently visible on the hillside as it finds its way through many ill-defined paths among boulders and grassy banks to fall at last as a trickling curtain into a lower pool in the valley.

The hillside from its crest to the valley stream will appear to have been cleared in the preceding decade according to the ancient tropical tradition of "slash and burn" by itinerant agriculturists. Tall rank grasses will overgrow small abandoned clearings, half-hiding ruins of the fallen forest. Several great trees will have been subsequently overthrown, root and mossy bough exposed to lie broken across the boulder-strewn slope. Some will have fallen against companion trees and ramp upward from the grass and herb cover into the crowns of these tall forest relics. . . .

VEGETATION

Here, on the thin soil cover, the forest has never been dense or continuous, but several large spreading trees will have found pockets of deeper soil and now dominate the scene. On surrounding hillsides—outside the unseen dry moat—crowded seedlings of broad-leaved evergreens [and] . . . bamboos and Fatsia will represent a phalanx of forest colonizers beginning the invasion of the abandoned fields. In protected pockets above the far cliff face, accent plants will spread their broad banner leaves. The cliffs themselves will be stained by weathering and fleshed green by mosses and ferns along their upper side where frequent trickles and seeps stain their faces. Here and there curtains of thin-stemmed Akebia and evergreen Clematis will veil the ancient rock faces. Above the stream-cut cliffs the forest will canopy occasional overlooks from which visitors may view areas of the opposite hillside and the gorilla band foraging below the lush regenerating vegetation.

Visitors will approach the gorilla exhibit along curving paths overhung with large-leafed magnolias and feather-leafed Albizzia. At a point of emphasis, a cluster of palms will be in prominent view, planted in tubs buried into the slope but removed every winter into the greenhouse.

It is intended that the visitor be slightly disoriented by the twisting pathways and dense growth so that when he comes upon a view of the exhibit, it may catch him by surprise, as if by breaking into a clearing he found he was not alone. . . .

The gorilla exhibit scenario also included technical aspects and documented several different types of barriers, including a hidden dry moat at the rear of the exhibit space, with a thick sawdust base to soften the impact should a gorilla fall. This moat allowed unobstructed views out and beyond the exhibit area, so that landscape in the distance was borrowed as part of the general exhibit view.

Technical specifications for this moat, the only traditional and zoo-tested component in the new exhibit, were obtained by surveying other zoos that had outdoor exhibits for gorillas and chimpanzees. Its height and width having been proved effective by experience, the moat was the only element of no concern. Escape was more likely, it was feared, via those parts of the boundary formed by rock walls. Experienced local mountaineers were invited to test their skill. A few suspect hand and toe holds were revealed in these exercises and amended.

Virtually no experience was available, however, to determine the probability of success with the vegetation. In the early 1970s, Cincinnati Zoo had attempted to keep gorillas in a grassed enclosure planted with sod, but the animals, being animals, simply tore it all up and threw it at the visitors. Chester Zoo had kept chimpanzees and orangutans in a grassed enclosure but with no other vegetation and had no experience with gorillas. The absence of hard data was a source of much anxiety.

Promotional news items for a zoo are never as important as the comfort level of the animals. Hence, there was no ribbon-cutting ceremony, no fanfare, no visiting public, and no dignitary present on the day that the gorillas first ventured into their new exhibition habitat. Instead, just a few people that the gorillas knew well waited with hushed excitement. The animals had been immobilized and given complete physical examinations in the old Ape House several days before, then moved into the den area of the new exhibit. Grilled doorways were left open so that they could see into the new outdoor space. On an early summer day in 1978, two staff members especially well known and trusted by the gorillas settled down among the tall grasses. They were there to provide a reassuring presence for the animals: familiar faces in a world where everything else was unknown.

The first gorilla released into the new area was the dominant male, Kiki, a 450-pound mature animal, in prime health, possessed of immense strength. He stood for several moments at the open doorway, quietly surveying the scene, pursing his lips, taut with tension. Then he stepped out. As he placed his hand on the grasses, feeling their soft yielding texture for the first time, he flinched and pulled back. This massive and unimaginably strong animal, a bulk of dense muscle, was intimidated by grass. Walls of concrete and doors of steel had trammeled his life. Exposed in a gloomy cell for years, he had never known the simple luxury of lying amidst grasses and shrubs in the open sunshine, the light filtering through leaves to his eyes, the wind stirring his fur, the rain wetting his brow.

What was dampening his face now, as he stood at the edge of his new home, were beads of sweat. His tension was palpable, yet he again stepped out into this alien green and sunlit place. He ventured a few steps further, flinching once more and cowering for a moment as a raucous crow flew overhead, a ragged black threat in the unknown sky. But he persisted, slowly and quietly exploring and expanding his world. Within an hour he had strayed quite deeply, plucked some grasses, tasted some leaves. Within a week all members of the gorilla troupe had also been successfully introduced and settled into their new home.

Remarkable changes in the troupe's behavior now began to emerge. Ritualistic threats between the two adult males were muted. Sudden body rushes were no longer resolved by physical contact but by one of the animals simply moving away and out of sight. Extreme inertia gave way to exploration and play. The group was undeniably more relaxed. They appeared content.

It is flabbergasting what minute amounts of information were then known of gorillas. One of the unforeseen changes was that the hair on the top of the males' heads turned quite ginger. Sunshine had lightened it, like kids on summer break. Another was the enthusiasm and apparent delight with which Kiki clambered high into the trees, almost to the very top, and sat for many hours looking out to the distant views of mountains and lakes.

Such a sight was so unusual, so unexpected in a zoo, that visitors looking into the exhibition habitat would often complain that there was nothing there. In fact there were usually two adult gorillas sitting facing them among the thick vegetation, like black shadows, and one adult male gorilla at least fifty feet above their heads, high in a maple tree.

Accurate information about the life of gorillas in the wild has entered public knowledge only in very recent years, thanks initially to the tenacity and fortitude of George Schaller and Dian Fossey. Much of what people believed before Schaller's and Fossey's research in the 1960s and 70s had its source from depictions of gorillas in violent mode, even in natural history museum dioramas, and from exaggerated stories. Robert Garner, in his pioneering ventures to learn more about the natural habits and habitat of gorillas, had provided extensive evidence of gorillas' naturally shy character. He documented the fact that gorillas are expert climbers and explained that they did not build huts as some averred and would not attack without provocation; rather they were shy and timid, and did not tear open the chest cavity of their adversaries and drink the victim's blood. His studies were not widely distributed and much of the population would have had no access to them. More people were exposed to another source of information, the unrealistic accounts of Paul Du Chaillu, the self-proclaimed discoverer of the gorilla, in 1859.

Du Chaillu recounted tales of gorillas raping women and of his being attacked by maddened gorillas who flattened his rifle barrel with their teeth. A pugnacious anti-evolutionist, he attracted thousands to his public lectures (accompanied by a stuffed gorilla in an aggressive pose) but raised the ire of the scientific community. Thomas Huxley considered him to be no more than a hunter masquerading as a scientific naturalist. Considering their content there can be little doubt that he fabricated many of his accounts, and though he shot more than twenty gorillas on one expedition to Africa, he never objectively observed the animals or their behaviors. He described them only as "hellish dream creatures"—a portrait the gorillas could more accurately have applied to white men with guns (Raby, 1996).

It is surprising how long it has taken to learn to see our closest cousins objectively. Jane Goodall has had to devote her life to reveal the true nature of chimpanzees. Before her, virtually no information was available about an animal that shares 98.6 percent of its genetic material with humans. The pioneering long-term studies of gorillas in the wild by Fossey and Schaller, under equally difficult physical conditions, revealed pictures of gorillas and their behaviors and lifestyles quite different from the historical stories based on ignorance and fear. From these objective research projects we learned the truth: gorillas are family oriented, intelligent, peaceful animals. Fossey and Schaller had assisted Woodland Park Zoo staff and the Jones & Jones design team on the new exhibit. Their words of encouragement and their challenge to try new approaches, as much as any practical recommendations, were the greatest motivations for the zoo management and design teams. (Fossey would often announce at her public lectures, in response to the inevitable question from the audience as to whether or not gorillas should be kept in zoos, that in her opinion only Seattle's zoo had facilities suitable for them. Twenty years later, with zoo gorilla developments such as those at Atlanta, Dallas, San Diego, Orlando, Tampa, the Bronx, and Melbourne, Australia, her answer need no longer be so restrictive.)

After several seasons it was clear that the gorilla exhibition habitat was a success—for the animals, the keepers, and the visiting public. Zoo visitors who once had stood in the grimy corridor of the old ape house, passively gawking or mocking the animals with whoops and shuffling jumps, now stood in small clearings amid dense vegetation and did not shout or howl or, often, even talk, but occasionally whispered to each other, with wonder in their eyes.

GORILLA ON THE SIDEWALK

There was, however, one dramatic failure. The exhibit keeper, Violet Sunde, had noticed in the first springtime that Kiki, the dominant male gorilla, had for several days been sitting on the edge of the hidden moat at the rear of the exhibit space, studiously surveying the scene. There had been startling revelations in the new habitat area as to the intelligence of the gorillas and especially this particular male. Keepers, for example, would call the gorillas inside several times each day (all large animals at Woodland Park Zoo were conditioned to respond to some particular sound by entering their dens) and then go out to scatter dried fruits, sunflower seeds, and nuts among the grasses as well as hiding chopped fruits and fresh vegetables. It long puzzled them that Kiki, when let back outside, would make his way directly to the hiding places of his favorite snacks. The answer was revealing and astounding. Within the keeper's service area a television monitor for a closed-circuit camera allowed surveillance of the outside area, and Kiki had discovered that by sitting in a particular corner he could just manage to see this monitor, though at a very oblique angle, and could watch where the keepers were hiding the food items. That he was able to translate this skewed view on a tiny black-and-white screen into the three-dimensional complexity of the

Nina and her baby Zuri at the Woodland Park Zoo, Seattle, 1983. The first naturalistic exhibition habitat for gorillas opened at Woodland Park Zoo in 1978. For the first time, zoo gorillas had trees to climb, places to hide, a complex landscape to explore, and live vegetation to interact with. For many years, the animals had lived in a concrete cage at the zoo, their boredom relieved mainly by bouts of tension-relieving agressive behavior. In the naturalistic enclosure, their behavior soon changed greatly: they became relaxed and contented. The exhibit's naturalism also generated a dramatically positive change in zoo visitors' perceptions of gorillas. (Author photo.)

outside world and memorize the placements he had seen, tells us much about gorillas' perceptual abilities.

It is small wonder that staff were concerned about what was going on in Kiki's mind while he was quietly surveying the hidden moat for several days. All came to light when two visitors walked into the zoo administration building one quiet damp morning, to complain that they didn't like the gorilla being out in the public area. Similar reports, however, had been made before; the landscape-immersion experience and the illusion of an unbounded exhibit area had deluded several zoo visitors into believing there was no barrier between them and the animals. The situation was explained to this pair of visitors, as it had been on earlier occasions to others. This time, however, as the couple left the office, the man was overheard to mutter his concern to his wife about there being a gorilla on the sidewalk.

The dimensions of the hidden moat at the Woodland Park gorilla exhibit had proven successful in containing gorillas in other zoos. In this instance it failed, and for one important reason. Apart from including several large existing trees, the designers had added some young trees in the exhibit area. Kiki had pulled one of these, a hawthorn with a four-inch diameter trunk, out of the ground, roots and all, dropped it into the moat, and used it as a ladder to climb down, then propped it against the opposite wall to clamber up and out. On his exit the tree had fallen over, and he was unable to return. He sat behind a shrub and tried to make himself as inconspicuous as possible, which was not very. Veterinarian Jim Foster was able to sit with Kiki, keeping him quiet and relaxed and feeding him from a pail of fruits and vegetables, and eventually immobilized him with an injection so he could be carried back on a stretcher. To discourage further attempts, an electrified wire was laced along the inside edge of the moat wall. Kiki closely observed it, touched it once, and never went near it again.

LANDSCAPE ARCHITECTS MAKE THE DIFFERENCE

No other zoo attempted to build an outdoor naturalistic exhibition habitat for gorillas for many years. They waited cautiously before being convinced that this type of exhibition habitat could sustain the daily predations of a group of gorillas. One Midwest zoo, six years after the Woodland Park Zoo's gorilla exhibit opened, launched a fund-raising campaign for a multimillion-dollar gorilla exhibit. Though publicized as a "natural habitat" exhibit, the promotion brochure did mention that there would not, "of course," be any live vegetation in the animal area. In 1980, Jersey Zoo, in the English Channel Islands, built a functional but unexciting open-air exhibit, with a lawn adorned by sawed logs and with viewing places where visitors look down on the gorillas over a visible perimeter wall. When Warren Iliff became director of Dallas Zoo, a new gorilla exhibition habitat was one of the first projects he initiated. Opened in 1989, it raised the standards and was an example of good original thinking with a very clear philosophy and a sound conservation message. About the same time, Singapore Zoo built a naturalistic exhibit for gorillas, though, sadly, it had to be abandoned when the animals contracted an endemic and fatal tropical parasite. In 1988, Zoo Atlanta, as part of its metamorphosis under the direction of Terry Maple, dedicated $6 million and 10 percent of its zoo site to an especially inspiring exhibition habitat, designed by Jon Coe, which was the first to maintain gorillas in contiguous social groups. Two years later a fine example designed by Stuart Green was built at Melbourne Zoo, Australia. In the early 1990s, a good naturalistic exhibit for gorillas was built at San Diego Zoo and one of equal quality at Busch Gardens, Tampa. At last zoos had accepted, almost at the end of the twentieth century, that family groups of gorillas could be maintained in zoo exhibits with abundant live vegetation, in replicas of their natural habitat.

Jones & Jones designed the gorilla exhibits at Dallas, San Diego, and Busch Gardens. After their work at Woodland Park, the firm developed master plans in the 1980s for zoological parks as diverse as San Diego Zoo (where, ironically, some senior staff in the 1970s had originally been strident opponents of the concepts of bioclimatic zoning and landscape-immersion exhibits), Belize Zoo, and the Jerusalem Zoo. In 1992 Jones & Jones prepared the master plan for the Arizona-Sonora Desert Museum and in the late 1990s completed design work for the African savanna at Disney's Animal Kingdom.

Woodland Park's long-range plan was the first modern zoo plan, and maybe the first ever, to be prepared by landscape architects. David Streatfield, chairman of the Department of Landscape Architecture at the University of Washington, says that it "transformed the design of zoos, and considerably enriched their interpretive function" (Streatfield, 1995). Traditionally, and still in Europe, zoos have employed architects for their planning and design work. The difference between the two professions, though they sound alike, is immense and is important. Architects are trained to think in terms of structures and in the assembling of materials and the way that manufactured components fit together. They seek expression through buildings. Landscape architects, conversely, work with landforms, natural systems, climate, microhabitats, and vegetation. There are exceptions; some individuals can always think outside the box. Johnpaul Jones, for example, who designed the world's first exhibition habitat for gorillas, at Woodland Park, is a registered architect, with no formal training in landscape architecture. Generally, however, architects and landscape architects approach design problems in characteristically different ways, and the results of the two approaches are clearly exemplified in zoo design. Small wonder that both Heini Hediger and William Conway have described the architect as the most dangerous animal in the zoo, and that Peter Crowcroft, ex-director of Brookfield Zoo and of Taronga Zoo, included in the index to his book *The Zoo,* "architects, snide comments about." Of greater puzzlement is why, if they were so unhappy, zoo managers persisted for so long in using only architects. Fortunately, since Jones & Jones started designing zoo exhibits, landscape architects are now more likely to be hired for such work.

DESIGNING FOR REALISM

A delightful example of the characteristically different design approaches between architects and landscape architects is revealed in a 1981 Asian primates exhibit by Jones & Jones at Woodland Park Zoo. It is not simply an enclosure to display animals, but a clever replication of an actual habitat. High, overhanging clay banks, ostensibly undercut by floods, enclose a sweep of shallow river rapids scattered with large boulders. Selected viewing areas face west and uphill, so that afternoon sun highlights the spray from the rapids and throws the background vegetation into shadow, emphasizing the density and exaggerating the depth of the apparent forest that surrounds the view. This exhibition habitat is designed to create the illusion that the visitor, after following twisting pathways overhung with broadleaf evergreens, has stumbled upon the view by chance, coming upon the animals by good fortune rather than by design. It suggests that the animals had been foraging in the adjacent forest and found the sunny, open areas to be pleasant spots to pass the afternoon. It is a type of design approach foreign to those who seek architectonic solutions.

The realism of much of Jones & Jones's work at Woodland Park sometimes surprised the experts. When photos of the gorilla exhibit were sent to Dian Fossey by Johnpaul Jones, she wrote back to say that she had left them on her desk and returned home to find some of her field biology students perusing them, trying to decipher exactly where in the forest they had been taken and who the individual gorillas were. On another occasion, a photograph of the patas monkey exhibit in Woodland Park's savanna exhibit was selected by National Geographic for publication in one of their books about Africa, assuming it to be a photograph taken in the wild.

A new course was set by Jones & Jones' work at Woodland Park Zoo, and the most important landscape architecture firms now specializing in zoo design in the United States—notably Coe and Lee, the Portico Group, and Sherman Yañez Mikami—are run by people who spent time working in the Jones & Jones office in Seattle and cut their teeth on studies at Woodland Park Zoo. "Landscape immersion" has now become a byword in the zoo profession since its 1976 introduction at Woodland Park Zoo, but the concept has not progressed and in some ways has regressed. Several landscape architects have learned tricks of the trade to produce naturalistic exhibits, but they have not extended the philosophies. Brainstorm sessions in the early 1970s between Jones & Jones and Woodland Park Zoo staff, for instance, to find examples of natural features that could be used for zoo exhibits instead of brick walls or wooden fences, resulted in such ideas as flood-cut earth banks and fallen trees as barriers or displaying animals in forest clearings ostensibly created by floods or itinerant agriculturists. These same ideas are now repeated so frequently, however, that zoos should become concerned that they are presenting a universal image of wild animals only in damaged and degraded environments. New thinking, greater effort, and more creativity is needed.

In many ways, the breakthrough of landscape immersion at Seattle in the 1970s mirrors the events at Stellingen in 1907, when Hagenbeck introduced his panoramas. There was initial rejection and bitter denunciation by many zoo professionals in both instances, followed by wholesale duplication. But the followers often miss the point of the exercise. Many zoos are now mimicking landscape-immersion techniques with passionless exhibits that reveal no true understanding of or sympathy for the philosophies behind the concept. Most regrettably, it has become routine to plant all the natural vegetation out of reach of the animals, beyond moats or hot wires. The surroundings may look realistic to visitors, but are of no direct benefit to the animals. The original intent of landscape-immersion exhibits was that both people and animals would be immersed in the same type of landscape, with the understanding that part of the zoos' responsibility was to periodically replace damaged plants in animal areas. When the gorillas at Zoo Atlanta started destroying the vegetation in the new exhibition habitat the designers and horticulturists implored director Terry Maple to install protective electric wires. He responded, rightly, with the advice that they plant cheaper trees.

As "immersion" becomes a cliché in exhibit jargon, there is almost a cookie-cutter approach to exhibit design once more, as zoos revert to their old practice of parroting each other's works, producing copies of copies that become ever more degraded and ill-defined in the process. Even Woodland Park Zoo has fallen into the trap. Although Jon Coe has returned there to design a very good quality Northern Trails exhibit, nicely replicating an Alaskan habitat, that zoo has also recently implemented an Asian primates exhibit by the Design Consortium that demonstrates far too many examples of what should be avoided in zoo-exhibit design. The exhibit has been lushly and dramatically planted by curator of horticulture Sue Nicol, but this cannot cover such basic design flaws as looking down on top of the animals or viewing through square windows set into the usual abundance of illogically placed and resolutely fake rock walls. When medieval monks were given the task of copying Latin and Greek manuscripts, they usually did not understand the texts. The results, after centuries of this blind reproduction, were often nonsense. The beginnings of a similar process are underway with regard to the landscape-immersion philosophy of zoo design.

FUNDAMENTAL NEEDS, SUPERFICIAL CHANGES

The strong criticism of zoos that arose twenty or thirty years ago has largely been dispelled, at least in the United States. Better standards of veterinary care now prevail with new medicines and technologies; environmental enrichment programs; continual improvement of accreditation standards; installation of a professional ethics code (in the American Zoo Association); and achievement of some very successful breeding results have combined to produce very different zoos from those of the 1970s. Most especially, however, the old iron cages have largely disappeared from the public eye. Zoo visitors like the new open exhibits and the abundant greenery in the zoo. It was mainly the sight of wild animals in sterile, barred cages that spurred the public criticism that began to grow in the 1960s. All seems to be fine now that it looks as if the animals are in naturalistic enclosures.

But all is not well. The first signs of degradation of the landscape-immersion philosophy are becoming sadly evident as its standards are being lowered, its purposes ignored. Buffalo Zoo is not the only one to have built an animal exhibit (this one for gorillas) entirely of concrete and called it a "habitat"—it just happened to be a trend setter in this regard. A new exhibit at Salt Lake City's Hogle Zoo comprises five sawed tree stumps on a hillock of broken rocks and has been named "The Primate Forest." Even when examples of success surround them, it is not unusual for zoos to fail to reach for good design. An African savanna exhibit designed by Jones & Jones in the late 1980s now has a small-mammals exhibit at its center—cramped and crude, a deformity of lumpen rockwork—which the zoo proudly boasts as its own work.

If none of this is yet discomforting to the general public, however, it is only a matter of time. Moreover, if those complacently reassured visitors were to see behind the scenes of some of the new naturalistic exhibits, they would instantly realize that the nineteenth-century menagerie still exists. It remains the rule that most zoo animals spend a great part of each day (and all night) in barren cages, where screams and the clanging of steel panels reverberate at painfully high levels against the walls. This problem of uncomfortable noise levels in zoo service buildings has not received sufficient attention. Medical scientists have shown that prolonged exposure to excessive noise can induce loss of hearing and that it can also invoke psychological stress (Australian Academy of Science, 1976). Surveys of two years' admissions to a psychiatric hospital serving an area around Heathrow, for example, showed that a significantly higher number of admissions came from the noisiest part of the region, suggesting that people already under stress are especially vulnerable to noise. If this supposition is correct, the same is conceivably true for many zoo animals. Acoustic engineers measure reverberation time, which is the time it takes for a sound to fall away to inaudibility. In a furnished living room it is about half a second. Remove the drapes, carpets, and furnishings and the room's reverberation time will be extended, to about three-fourths of a second. That small difference, we know, is uncomfortable. People say that such spaces feel harsh. If our fairly inadequate perception of these variables is so sensitive, we should suspect that the aural environment may be of special importance to wild animals.

Zoos have given much emphasis to greening the public spaces in recent years, but behind the scenes, the typical zoo remains more Alcatraz than Arcadia. The very conditions Hediger eloquently argued against more than fifty years ago are still the norm; they are now just out of sight.

Zoos have the highest reason for seeking excellence in everything that they do, in the public eye or not. Food services, standards of restroom hygiene, grounds maintenance, gift shop products, education programs, employee morale and training, and advertising and promotion projects, all combine to create an experience for visitors that directly affects their attitudes and subsequent commitment to wildlife.

Hopes and ambitions by zoo supporters have often been too low. H. J. Massingham, writing in the 1930s of the future for London Zoo, hoped to see the day when "We shall see every iron bar in the Gardens painted a lively green." More recently, Allen Nyhuis (1994) has made the giddy claim that Busch Gardens's "animal exhibits, shows, shopping bazaars, restaurants, and over twenty thrill rides create a believable African atmosphere." Robert Halmi (1975) described Thoiry Game Park, France (which specializes in cross-breeding tigers and lions), where animals are displayed in flat paddocks devoid of grass or trees, as a place where people can see giraffes "in their natural environment."

There is a strong propensity among British zoos in particular, with some other European zoos in close agreement, to claim that if the exhibit space has met the needs of the animals, then it cannot (or need not) be criticized. Their defense is routinely accompanied by a complaint that they don't have the same level of funding as American zoos. Very significant differences in general standards of, and specific national attitudes toward, British and

Holding cages at zoos, in which animals typically spend most of their time, are reminiscent of the menageries of the nineteenth century. (Photo: Ian Smith.)

American zoos are illuminated by the entries on zoos in the British and American versions of the *Encarta* encyclopedia. The 1997 British version states: "Concern has been expressed for animals kept in zoos . . . owing to the unsuitable housing, lack of space, and unnatural habitat and climate in which the animals are often kept. These conditions have been suggested to promote abnormal behavior, such as pacing. Zoos maintain that they provide educational, zoological and conservational benefits." The U.S. version conversely describes zoos as places "in which live animals are kept for public recreation, education, and conservation programs. Modern zoos offer veterinary facilities, provide opportunities for threatened species to breed in captivity, and usually build environments that simulate the native habitats of the animals in their care."

The crude-looking monkey and ape exhibits at such British zoos as Twycross and Howlett's are actually praised by English zoo apologists. Jordan and Ormrod (1978) describe the placement of a tree stump in an ape exhibit at Twycross as "exquisite attention to detail." Howlett's ape cages are typically defended by the assurance that they provide vertical and horizontal surfaces for climbing. In other words, they are cages. They are certainly more useful for an ape or monkey than an open pen with a dead tree stump in it, but is meeting the animals' needs the only reason and purpose for building a zoo exhibit? Clearly that is the first essential, but seamlessly attached to that come the important needs of the human visitors and the justification for displaying animals at all. Zoos are built primarily and essentially for people. A refuge built principally for animals would be quite a different place, with far more space and quiet.

EXPERIENCING REALISM

The most compelling and obvious impact on visitor attitudes toward wildlife is the way that zoo animals are presented. This is why quality of exhibit design is of paramount importance. The validity of the zoo experience hinges on the functional *and* visual integrity of the zoo exhibits. Thus, zoo-exhibit design must flow from clear and strong philosophies (of Nature and of design) and be based on recognition of the realities and complexities of wild habitats and the extent to which human development and personal fulfillment relies upon the natural world.

Quality of environment directly affects the perception of zoo visitors, shaping their attitudes toward wildlife, and there seems to be little doubt that much of this is subliminal. Zoologist Amanda Embury (1992) recounts a survey at Melbourne Zoo in which visitors were asked to select from a list of adjectives to describe gorillas. The first part of the 1988 survey documented responses to seeing gorillas in an exhibit that was essentially a concrete pit and typical of most zoo exhibits until recent times. Predominantly, people chose negative words—"vicious," "ugly," "boring," and "stupid." Two years later, after the gorillas had been relocated to a large, new naturalistic exhibition habitat that replicated the African rain forest, visitors had completely opposite responses, selecting adjectives such as "fascinating," "peaceful," "fantastic," and "powerful." Unless they are being exposed to extremes of horror or wonder many zoo visitors assess the exhibits and their own relationships to the designed environments only subconsciously. It is for this reason, ironically, that attention to precise detail is essential.

The presence of professionally trained designers on staff could help to ensure more consistently high standards. Two or three decades ago, it was unusual to find trained educators, marketing personnel, or even veterinarians as zoo staff members. Major zoos today consider such expertise essential for their operations. It is overdue for zoos to recognize that they need strong and stable creative-design teams as an integral part of the staff. Just as a zoo director or curator cannot be expected to have expert knowledge about learning techniques, matters of animal health care, or marketing, so they should absolve themselves of the responsibility for conceiving exhibit designs.

The plea for better design in zoos must not be mistaken as a concern only for aesthetics. There is a dimension in zoo design that goes beyond such considerations as visual balance, massing, harmony, and the integrity of materials. Zoo environments must also, by design, exemplify attitudes of respect for Nature. The forms and features resulting from application of good design principles will inevitably differ from zoo to zoo, depending upon funding, climate, site restraints, local emphases, and a multitude of other factors. Entry plazas, restaurants, service buildings, rain and shade shelters, and all other built forms for people can legitimately reflect local traditions, climates, and materials. But in the animal exhibit areas there must be one constant and inherent design philosophy: Nature is the norm.

Henry Beston, spending a year alone on the farthermost peninsula of Cape Cod in the 1920s, became intimately tuned to the details and complexities of the ocean, beach, sky, and animals. While discovering the delights of the sounds, scents, colors, and shifting moods of this wild and lonely place, he found the words to express, in his book *The Outermost House* (1928), a powerful statement about Nature that perfectly encapsulates what zoos should aim for:

> We need another and a wiser and perhaps a more mystical concept of animals. Remote from universal nature, and living by complicated artifice, man in civilization surveys the creature through the glass of his knowledge

> and sees thereby a feather magnified and the whole image in distortion. We patronize them for their incompleteness, for their tragic fate of having taken form so far below ourselves. And therein do we err. For the animal shall not be measured by man. In a world older and more complete than ours they move finished and complete, gifted with extensions of the senses we have lost or never attained, living by voices we shall never hear. They are not our brethren, they are not our underlings; they are other nations, caught with ourselves in the net of life and time, fellow prisoners of the splendor and travail of the Earth.

Zoos are a metaphor for our attitudes to and relationships with Nature. The critical importance of landscape immersion as a technique for zoo design is that it acknowledges, makes evident even, the importance and the values of

An Indian rhinoceros in its natural habitat (facing page) is undeniably impressive, the scene almost reminiscent of something primordial. But an Indian rhinoceros on display at the Philadelphia Zoo in the 1980s (above) is merely pathetic, its inherent majesty so reduced by the paucity and ugliness of its environment as to make the animal a clumsy caricature of itself. (Photo on facing page © James Foster; author photo above.)

natural systems. It creates opportunities for zoo visitors to experience something more meaningful than passively looking at an animal on display. Moreover, careful application of landscape-immersion philosophy, with attention to concealed barriers and authentically replicated forms of the natural landscape and use of borrowed vistas and studied sight lines, can all combine to create a memorably evocative experience in which zoo visitors associate wild animals with appropriate wild habitats. It achieves two important goals. Zoo visitors, even if they don't read the interpretive graphics, can learn by

A Celebes macaque stretches an arm forlornly to the world outside its sterile cage at the Paignton Zoo, England. It would be easy to misinterpret this as a mute appeal for help and consolation from its enforced isolation, because we don't know just what the animal is thinking. Clearly, though, such conditions are not conducive to natural behaviors, and it is safe to conclude that this must be an unhappy and stressed animal. Moreover, the conditions in which it is displayed could never elicit any sense of admiration or wonder for what is in fact an admirable and wonderful being. (Author photo.)

associative intuition that certain animals and certain habitats are inextricable. And they can by similar association gain more respect for wildlife. A wild animal seen in the context of its habitat carries a natural dignity. The more degrees of distortion in the representation of that habitat, the more unnatural the animal both behaves and appears.

CHAPTER SEVEN

AGENTS OF CONSERVATION

Few people have heard of the pig-footed bandicoot or the great amakihi. Certainly no one will ever see them alive. These animals became extinct in the first decade of the twentieth century. Should it be of any concern that we will never know anything of the huia or the black mamo? Were they just inconsequential creatures, saddled with strange names, like the snail-darter? Does it matter that our children will never watch an orange-breasted flower-pecker or an African forest shrew? These small animals, in remote corners of the world, have left us forever. Specks of life and light, they have been extinguished, leaving the world a little darker, a little colder.

For short-term commercial gain, humankind continues to eradicate wild species of plants and animals. Biologist Edward O. Wilson (1989) calculates that Earth is losing species, principally in tropical rain forests, at the rate of six an hour. Fewer than 3 million square miles of tropical rain forest exists, and if its destruction continues at the same rate as it did between 1979 and 1989, the last tropical rain forest tree will fall in 2045. This rate, however, is increasing (Conway, 1999). Almost all of the big cats, the antelopes, wild horses, great sea turtles, all the storks and cranes, the vast majority of the world's frogs, crocodilians, and migratory birds, all the rhinoceroses from Africa to Sumatra, the gorillas, and unknown numbers of unnamed little creatures are in decline. Some very small creatures exist in such extreme and isolated microhabitats that their entire population can disappear when a portion of ancient forest is clear-cut. Even the populations of some very large animals are in free fall. Addaxes tumbled from five thousand in 1966 to fewer than five hundred by 1990 (Correll, 1994). The black rhino population declined by 87 percent between 1986 and 1996 (Fouraker and Wagener, 1996). At the beginning of the twentieth century, there were an estimated eighty thousand tigers in India. Seven decades of big game hunting reduced that population to fewer than two thousand (Marr, 2000). Hunters continue to take many millions of animals out of the tropical forests each year. An editorial in a recent issue of *Oryx,* the International Journal of Conservation, (Robinson, 1999) explains that perhaps 1 million tons of wild animal meat is now taken annually from the forests of the Congo Basin alone. These are statistics beyond comprehension and portend a calamitous result.

Most zoos and many books and magazine articles about zoos in recent years have responded by claiming, repeatedly and loudly, that captive-breeding programs for wild animals make zoos the modern Noah's Ark. This simple imagery is ludicrous. We cannot save the world's endangered wildlife through the few successful breeding programs in zoos, just as one cannot save a language simply by holding on to a rare document.

Loud self-congratulation by zoos about the contributions they are making to conservation booms noisily, but it is amplified within a relatively empty barrel. Edinburgh Zoo recently produced a brochure with the casually sweeping declaration that during the past century "Zoos . . . progressed from centers of entertainment to centers of conservation." The statement does not withstand scrutiny. William Conway (1995c) more accurately suggests that "the future of zoos is to *become* Conservation Parks" (my emphasis) and points out that making that transition "is a high step, and the footing is slippery." The climb, moreover, is not up to a simple peak, but over a mountainous range. A conservation park would be more than just a sanctuary for rare animals or a place that reproduces endangered species. It would also be actively engaged in effective programs for saving wild habitats. It would employ researchers, writers, photographers, publicists, fundraisers, biologists, wildlife managers, lawyers, ecologists, and others in the cause of conservation, and it would work in wilderness areas and be directly involved in saving threatened habitats. It would campaign, too, for effective legislation to protect wild places, to stop pollution, to control exploitation of natural resources. Its programs and exhibits would enthrall and delight people, so as to convert them to the cause of conservation. Most especially, a conservation park would dedicate itself to calls for action and give the public explicit guidance in ways to save wilderness and the natural environment.

The World Zoo Organization's Conservation Strategy exhorts zoos to "make as great a contribution as possible to protecting holistic life systems and to conservation education." Such a task is undeniably enormous. Convincing people to develop a new respect for Nature, to make changes to their lifestyle, and to adopt different values will be laborious. It might not even be achievable. In China, for example, where the air in Beijing and many other cities is so polluted with coal dust and fumes that often the sun cannot be seen through the murky gloom (one city, Benxi, even disappeared from satellite photos under its pall of filthy air), and where more than one in four deaths is caused by respiratory disease (Hertsgaard, 1997), people seem willing to tolerate life in a land deforested first by Mao Zedong's "Great Leap Forward," which included the cutting of millions of trees, and more recently by acid rain, and with rivers and lakes poisoned by untreated factory effluents, breathing air thick and toxic, all for the sake of more jobs and better pay. In more economically advanced countries, meanwhile, we continue to consume the world's natural resources at an unsustainable rate, for similar short-term gain.

SURVIVAL PLANS

Issues such as economic advancement should be approached with caution, but zoos can immediately stop degrading the word "conservation" by employing it so irresponsibly. As one example, the December 1995 issue of the American Zoo Association's newsletter announced the opening of a "Chimpanzee Conservation Center" in a Midwestern zoo. The title implies a place focused on matters of chimpanzee conservation. In truth, it is merely another new zoo exhibit whose title is its only impressive feature. Consisting of a thousand-square-foot brick building, it comprises two holding cages and an "exercise area" furnished, the newsletter reports, with "ropes and a swing."

Claiming that such a facility is a conservation center trivializes the whole concept of conservancy. Similarly, when an endangered species is born, zoos rush to produce press releases about their latest contribution to conservation. Their enthusiasm over the birth of a rare animal, however, is no excuse for damaging overstatements.

Most people have such scant contact with the natural world that they do not yet directly see or feel the destruction and pollution of wilderness; the blast holes in the web of Nature are still beyond their personal horizon. It is comforting for them to believe that the local zoo or botanical garden is saving the world's endangered species. But all this creates a false sense of public complacency. The American Zoo Association publishes a pamphlet, *The New Ark,* claiming that "The survival of the world's endangered wildlife pivots on the conservation and education efforts of modern zoos and aquariums." If it were true, this would be a terrifying prospect. Zoos deal with only a tiny fraction of the world's fauna. Moreover, they concentrate almost entirely on breeding the species that are important for zoos, not those important to Nature.

There are about forty-six thousand vertebrate species in existence. The most optimistic projections state that if all the world's professionally operated zoos, in concert and under perfect conditions, devoted a full half of their facilities to breeding endangered animals they could perhaps manage about eight hundred of them in viable breeding populations (Conway, 1995a). By contrast, the 1995 Global Diversity Assessment published by the United Nations cites fifty-four hundred animal species threatened with extinction. In the past four hundred years, about five hundred animal species have been lost, with much higher figures for plants, at an extinction rate at least one hundred times greater than would be expected without human presence (Groombridge and Jenkins, 2000). Most people remain unaware of the scale of this catastrophe. The general public knows only about those glamorous few animal species that receive publicity from zoos claiming to save the world's wildlife—such as tigers, gorillas, California condors, and rhinoceroses. In reality, these are only the tip of a massive berg of life forms sinking into extinction.

The fact that zoos, by their very nomenclature, are restricted to zoology is a potentially fatal flaw. Centuries ago, Europeans made sense of the apparent chaos of Nature by an analytical, systematic approach, dividing it into kingdoms and classes and subdivisions. Disparate natural history institutions continue to reflect that system; almost any city has a zoological park, a botanical garden, and a natural history museum. If, however, the story of Nature is to be lucidly told, it needs something other than this patchwork of natural history institutions. It is a story of connections that cannot be understood in a piecemeal presentation; a new type of institution is needed. Zoos may not be readily able to change their name, but they must become quite different in content and program. They must metamorphose.

THE BIGGEST PROBLEM

The World Resources Institute, the World Conservation Union, and the United Nations Environmental Programme (1995) have all declared the loss of biological diversity to be the biggest conservation problem we face.

Only a handful of biologists have brought attention to the dangers of the loss of biodiversity: principally, Tom Lovejoy of the Smithsonian Institution, Norman Myers of Oxford, Tom Eisner of Cornell, and E. O. Wilson of Harvard, who is often publicly acknowledged as the father of biodiversity. These scientists have become directly involved in the political process, testifying to Congress and, in Lovejoy's case, personally escorting politicians into the planet's richest regions of biological diversity, the tropical rain forests. At least 10 percent of the U.S. Senate, several members of the House of Representatives, Al Gore, and many others have, thanks to Lovejoy, become more familiar with this rich habitat (as well as being exposed to his infectious fervor for preservation of Nature).

The term "biodiversity" seems to have made its first formal appearance in 1986, at a U.S. National Forum on Diversity in Danger, cosponsored by the National Academy of Sciences and the Smithsonian Institution. It is a shorthand phrase for a complex notion and was invented to help people generate a new view of Nature and its value to us and to encourage a change from thinking of preservation of species to conservation of wildlife.

"Wildlife," however, does not generate universal appeal. Undeniably, we are drawn to life forms that fit our cultural bias, fascinated by plants and animals that are showy, colorful, pretty, gigantic, cute, or freakish. Traditionally, we have decided the value of wild places by aesthetic preferences. It was not especially difficult to persuade politicians to protect Yellowstone, Yosemite, Mount Rainier, the Grand Canyon, Bryce, or Zion as national parks. But it was not until 1947 that a swamp habitat in the Everglades was given the same protection. No grassland in North America has yet been elevated to this status.

Madagascar is one of the most severely damaged places on Earth and is home to some of the rarest species. Isolated for 70 million years, the processes of natural selection have been compressed and accelerated here, resulting in the evolution of numerous endemic species. More than 90 percent of Madagascar's plants and animals occur nowhere else on the planet. But of all these strange and endangered life forms, most attention has been given to saving one group of furry animals, the lemurs. Ring-tailed lemurs, with their dove-gray fur, amber-yellow eyes, and handsome striped tails are special favorites in zoos. Truly, if this physically impressive animal could be employed as an ambassador for the whole island of Madagascar, then the claim by zoos that their focus on charismatic vertebrates is encouraging the preservation of wild habitats would have some validity. It has the potential to be a very effective idea, but it will require much more than simply exhibiting the species to zoo visitors. Display is only the first step. It must be supported by programs that aim to create a better informed, more energized, and more caring public.

In the early 1980s it became clear to many conservationists that trying to save animal species after they had reached endangered status was not likely to be successful. Many of the endangered species were integral components of a complex set of relationships with other species, both plant and animal. It was not *species* that needed attention, but preservation of whole and viable *habitats,* of biological diversity.

Of all species that have ever lived on Earth, 99 percent are now extinct. Extinction is an inescapable part of the evolutionary process, but extinctions today are following a pattern never seen before. Norman Myers, a British ecologist working in Africa, was probably the first, in the mid-1970s, notes Parsons (1996), to try and quantify the extinction rate of the modern age. He calculated that between 1600 and 1900, humans had caused the loss of about seventy-five species of wildlife. Another seventy-five of the known species then disappeared in little more than the next half century. But Myers also calculated that maybe twenty-five thousand plant species and hundreds of thousands of invertebrates were in danger of extinction. He projected a potential extermination of millions of species. For this audacious bell ringing he was soundly criticized, especially by academic economists. Now, with better data, we know that Myers was essentially correct.

In spite of enormous brainpower, humans are sometimes indescribably stupid. Because of the unhygienic ways of farming ducks and pigs in Asia, there is risk of a new influenza pandemic. In pursuit of political gain, armies defoliated forests in Vietnam and bombed the fields of Flanders into seas of mud and blood. The precious gift of antibiotics has been squandered, and soon there may be none that are effective to protect our children. Industries have fouled our wetlands and rivers with chemicals, fished oceans dry, and gouged holes in the earth's protective ozone layer. And though many commonly profess to be animal lovers (a term which usually only means an affinity for cats or dogs), humans have been wiping out animals at an ever increasing rate over the past century. This litany of doom and idiocy has reached terrifying proportions.

Moreover, we are now exterminating not just animal species but those natural environments in which species adapt and evolve over time. Wild habitats are being poisoned and eradicated. There have been five major implosions of loss of species in the earth's history. The sixth, human caused, is now underway. Zoos must therefore not restrict their conservation efforts to breeding orangutans and jaguars and rhinoceroses, laudable as those efforts are, but must involve themselves in bringing vital messages to people about the need to save wild habitats. In this way, humankind can save homes not just for impressive mammals and colorful birds, but also for all the tiny organisms that produce oxygen, that fix nitrogen, disperse plant seeds, pollinate fruit trees, and wage biological war on plant pests: in short, that make the planet comfortably habitable for humans.

Zoos have been very effective in publicizing their breeding efforts. A 1995 survey by the Roper Organization, commissioned by Sea World, revealed very positive attitudes toward zoos in the United States, showing that a majority of Americans believe "zoos play a unique role in . . . conserving the environment" and that the most important functions of zoological parks are "the preservation of wildlife . . . and the prevention of extinction." The zoo-going public, however, may not be the best to make judgments on such matters. A nationwide survey by Stephen Kellert (1981) of three thousand Americans identified nine basic groups of attitudes toward animals. The zoo-goers group revealed considerable affection and concern for wild animals, but had one of the most limited levels of knowledge and understanding about ecological issues.

The concept of biodiversity has taken a long time to enter public consciousness. A 1994 poll by Louis Harris and Associates showed Americans to be "clearly unaware of some of the most fundamental facts" about biodiversity and incapable of identifying any cause for its diminution: only 8 percent were aware that habitat destruction caused reduction in biological diversity. In another survey, the Communications Consortium (1994) found that only one in five Americans claimed to have even heard of biodiversity, and Nabhan (1995) notes that in a Defenders of Wildlife survey *zero* percent of the fifteen hundred people interviewed volunteered loss of biological diversity as a conservation problem. Things are no better on the other side of the Atlantic. A 1995 survey in Britain asking people what they understood about biodiversity found that a majority thought it was a type of washing powder.

THE SANCTUARY

The species that have been saved from extinction through zoo captive breeding programs are notably the Mongolian wild horse, Père David's deer, European bison, Arabian oryx, Hawaiian goose, and the golden lion tamarin. The preservation of at least two of the species on this short list, namely the Mongolian wild horse and Père David's deer, was merely a fortuitous accident of history. The duke of Bedford, president of the Zoological Society of London, set out to obtain a herd of Mongolian wild horses in the late nineteenth century, since they were becoming rare. His expedition captured fifty foals, by shooting their mothers and the herd stallions. If the young horses had not happened to breed in captivity on the duke's estate at Woburn, the species' demise would have been hastened. The duke also had a very small herd of Père David's deer in his collection. This species became extinct in the wild about two thousand years ago but survived in the Imperial Hunting Park, near Beijing. In 1895 a flood breached the perimeter wall and starving peasants raided the park and killed most of the deer. The few that remained were slaughtered five years later, during the Boxer Rebellion. It was pure chance that a few representatives were kept at Woburn, that three of them bred, and that the species survived.

In any case, perhaps one should say that a few determined individuals, rather than zoos, have helped to save some species from extinction, such as Peter Scott with the Hawaiian goose at the Wildfowl and Wetlands Trust in England, Devra Kleiman's pioneering work with the golden lion tamarin at the National Zoo in Washington, D.C., and the persistently focused efforts of Gerald Durrell at his zoo on the Island of Jersey, specializing in species ignored in mainstream zoos.

Until surprisingly recent times, zoos bred their animals in haphazard fashion, not minding which male fertilized which female as long as baby animals greeted springtime visitors. In 1979 Katherine Ralls, a researcher at the National Zoo, working with Kristin Brugger and Jonathon Ballou, published the results of their studies on juvenile mortality rates as a correlation to inbreeding in sixteen species of animals that had been born at the zoo; species as diverse as deer and giraffe, zebra and hippopotamus. The facts were suddenly stark. The death rate of young inbred animals was markedly higher than those born from unrelated parents. A follow-up study on forty-four species confirmed the results. It was immediately apparent that zoo populations had to be governed in a different way. Thus, in the early 1980s, the American Zoo Association's conservation mission was defined as "intensive population management," by breeding animals in genetically regulated programs.

Worldwide, coordinated breeding efforts are now concentrated on more than one hundred species. Some zoos now employ biologists, geneticists, and reproductive physiologists to propagate rare and endangered species. Management programs for a wide range of captive populations, with studbooks and specific breeding strategies, ensure maximum genetic variation. All of these are commendable and valuable activities, but they are not sufficient. Zoos, with their unique skills, public voice, and global connections, are capable of much more, particularly in developing quality exhibits and education programs to change public attitudes and in efforts to save wild animals in their native habitats.

SURVIVAL OF THE SPECIES

The American Zoo Association coordinates a Species Survival Plan (SSP) that concentrates on comprehensive breeding programs for about one hundred thirty species. (There are also the European Endangered Species Programme [EESP], Indian Endangered Species Breeding Program [IESBP], Australasian Species Management Program [ASMP], Species Survival Committee Japan [SSCJ], and other similar programs around the world.)

SSP is more accurately an acronym for a Self Supporting Program for zoos. Not that this is necessarily wrong. It makes good business sense, for one thing. But for honesty's sake, if not other considerations, it should be recognized that most zoo breeding programs are aimed more toward conservation of zoo collections than to conservation of the world's endangered wildlife. The idea of zoos as latter-day Noah's Arks has much appeal and is seductive; it has beguiled even the most intelligent analysts. British science writer Colin Tudge (1991) said, "If zoos did not exist, then any sensible conservation policy would lead inevitably to their creation." But, sadly, all this, as National Zoo director Michael Robinson responded (1993), is "the stuff of myth." Any sensible conservation policy would be based on ecological needs and would ignore many zoo species. It would also recognize that zoos are not necessarily the best places for breeding rare and endangered animals. For example, zoos tend to select animals suitable for zoos and then opt for breeding the docile, tractable specimens. A difficult or aggressive animal is not likely to be selected for a zoo breeding program, yet it might have the very genetic characteristics that could best equip its progeny for survival in the wild. Moreover, zoos are for display and education: they cannot maintain the large numbers of individuals that are needed, and their exhibition facilities are not necessarily ideal for effective breeding programs, which are better undertaken on large tracts of rural land, such as the Bronx Zoo's facilities at St. Catherine's Island, in Georgia, or the National Zoo's off-site breeding property at Front Royal, Virginia. At such sites there is no competition for expensive exhibit space, better conditions can be maintained for hands-off management, husbandry regimes can replicate the animals' natural activity cycles, and individuals can be conditioned for the least stressful reintroduction to the wild.

Understandably, zoos concentrate on breeding the endangered species in their collections, but it must be understood that this has nothing to do with conservation of biological diversity. In the 1990s, zoos started to talk of their new role in saving biodiversity, as if it were a thing, like a tiger, that can be saved and be put on display. They also seem to promote the notion that biodiversity has to be exotic. In *The Origin of Species,* however, Charles Darwin contemplated the beauty of the diversity to be found in the commonplace English hedgerow: "It is interesting to contemplate an entangled bank, clothed with many plants of many kinds, with birds singing on the bushes, with various insects flitting about, and with worms crawling through the damp earth, and to reflect that these elaborately constructed forms, so different from each other, and dependent on each other in so complex a manner, have all been produced by laws acting around us." There is, he concluded, "grandeur in this view of life."

Zoo breeding programs have undeniably helped save some species from extinction, but they cannot save habitat and captive breeding cannot save animal populations. The endangered species in our zoos are rare cargo, indeed, and should be treated as treasures, but they are no substitute, as Conway (1995c) says, for "the loss of those wondrously rich and populous wildlife communities now dying with the tall trees of the great rain forests; for the disappearance of the awesome colonies of seashore pinnipeds and seabirds; the great herds of migratory ungulates in their thousands—for all those grand wildlife spectacles that so move our hearts and inspire our souls."

GETTING INVOLVED

At one time, zoos were merely collections of animals, like jewels, for the wealthy. There is now a growing sense of responsibility for presenting zoo animals as a means of study and enjoyment for an increasingly urbanized populace, and of creating facsimiles of wild environments that resemble their models in appearance, mood, and content. Only a brief review of the history of zoos makes it plain that we cannot tolerate any reversion to zoos as collections of living objects plundered from the wild. There is a very different need for zoos today and going to a zoo to see pandas and tigers is no longer sufficient justification for its existence. The wild homes of those cherished beings are disappearing, destroyed by humans. Zoos can and must become gateways to the wild, metaphorically and practically.

Happily, there is a growing awareness that zoos must become directly involved in wildlife habitat protection (what conservation biologists call in situ conservation). Frankfurt Zoological Society and Gerald Durrell's Jersey Wildlife Preservation Trust were pioneers in this area of conservation activity. Today, Minnesota Zoo has formed a partnership with Ujung Kulon National Park in Indonesia to help save the habitat of the Javan rhinoceros. Brookfield Zoo, under the leaderships of Peter Crowcroft and George Rabb, has given support to conservation areas in South Australia since 1971 (it currently focuses on the Bookmark Biosphere Reserve, which has emerged as one of the most promising models of community-driven conservation in existence) and gives grants to conservation projects in seventeen countries outside the United States. The National Zoo's Chris Wemmer has established international training programs. Fort Wayne Zoo has an education and fund-raising program to help conserve endangered monkeys in the Mentawai Islands of Indonesia. The American Association of Zoo Keepers has raised funds for The Nature Conservancy to purchase wildlife habitat in Costa Rica.

The conservation programs of the Bronx Zoo are the most extraordinary. Since its inception at the end of the nineteenth century, this zoo has been directly involved in conservation projects across America's wild lands. In 1956, it expanded its focus to include Africa, sponsoring ecological studies across that vast continent, training native wardens, funding antipoaching programs, and helping create or enlarge new wildlife parks. Today the Wildlife Conservation Society (which in 1993 changed its name from the New York Zoological Society) employs sixty scientists and in excess of one hundred research fellows in more than three hundred wildlife conservation projects in fifty-two countries and is engaged in environmental education programs in forty-nine of the fifty United States (South Dakota is the single exception). It is now responsible for the Wildlife Management Program for the nation of Malaysia. It has initiated wildlife education programs in Papua New Guinea and in Yunnan Province and is working with the Chinese Government to set aside one hundred thousand square miles to be protected in Tibet. Society staff are coordinating a seven-nation effort to create a protected wildlife corridor from Belize and Guatemala to Panama. They have formed a productive working relationship with Amboseli Park in Kenya. Within the past decade the society has been responsible for setting up 130 million acres of wildlife reserves across the planet and has recently helped establish Amana reserve in Brazil, the world's largest rain forest sanctuary, covering 22,500 square miles.

As if these specific conservation programs were not impressive enough, the staff at the Bronx Zoo extend their activities beyond the animals in their direct care. The zoo's veterinarians travel the globe to assess wildlife health, develop monitoring techniques, and train local people; the curators provide expertise in captive breeding of endangered species in sanctuaries and zoos across Africa and Asia; education personnel host teacher-training workshops throughout North America and around the world; science staff help researchers in the field with computer mapping and genetic analysis; and the design staff have assisted by creating master plans and exhibits for zoos in Entebbe and Nairobi, where increasingly urbanized populations are rapidly losing contact with wildlife.

More than twenty years ago a proposal was made to (and rejected by) the AZA executive director that each accredited zoo should be encouraged, or required through the accreditation program, to select at least one wild place, type of habitat, or community of threatened species, and adopt it. Each zoo could then fund studies, organize visits, make connections with local officials, discover local conservationists, and publicize their problems and their successes. If every major zoo made such investments in conservation around the world, there would be a vast new hope. In practical terms, zoos would become directly involved in funding, training, and offering hands-on expertise to the local biologists, wildlife managers, veterinarians, antipoaching squads, geneticists, scientists, publicists, demographers, and all other conservationists in their selected area. In return they would make valuable personal contacts and gain direct knowledge and information about

their adopted areas, habitats, and animals. All over the world, zoos need to get involved in similar hands-on activities while also trying to change public attitudes in support of wildlife conservation in their own region.

CALLS FOR ACTION

Debate on the threats to humanity posed by the massive and widespread loss of the planet's biological diversity, the damage to the environment, and the depletion of natural resources has centered on the economic consequences. These are not to be ignored, but there is another consideration that zoos can draw attention to: the value of biological diversity for our emotional, intellectual, and spiritual well-being. Humans surely depend upon a vast matrix of affiliations with other living things to achieve lives that are rich in meaning. Zoos can play a lead role in presenting that message.

Two of the most significant flag bearers of zoo conservation programs in modern times: William Conway (facing page), recently retired president of the Wildlife Conservation Society who started as curator of birds at the Bronx Zoo in 1956, and Gerald Durrell (above), who founded the Jersey Zoo in 1959 and, three years later, the Jersey Wildlife Preservation Trust. Their efforts in wildlife conservation serve as inspiration for other zoos. (Conway photo: © Wildlife Conservation Society, headquartered at the Bronx Zoo; Durrell photo: Nic Barlow, courtesy of the Jersey Wildlife Persevation Trust.)

Zoos are rightly proud of the numbers of people who visit them: in North America, zoo attendance statistics exceed all professional sports events combined. There is enormous potential to get good, basic conservation information to an audience that cuts across all socio-economic lines and all age groups and to help change public attitudes. It would be enormously beneficial if zoos began to help their visitors understand such phenomena as the dynamics of ecological systems and why biological diversity is vital to the health of our planet. This may well mean proselytizing, and many zoos are going to be loath to do this, but effective conservation will need a citizenry that is strongly sympathetic to the cause, energized about and supportive of the policies and politics that will protect wilderness, and willing to forgo some natural resources in favor of saving wild habitats.

As a powerful advocate for conservation, zoos could serve as Nature's conscience and ambassador. Better yet, they could strengthen this position by asking others to join their efforts. People need to understand that Nature is made up of complex ecological relationships, not just a series of interchangeable parts. To tell this story, zoos would have to explain the value of whole and fully functioning ecosystems and the interconnections of natural processes. It is very like what Fairfield Osborn was advocating at the Bronx Zoo more than half a century ago with his concerns about loss of natural resources, and it means that zoos must begin to embrace disciplines other than just zoology. To be good conservationists and effective interpreters of Nature, zoos must get involved with botany, geology, paleontology, ecology, and a host of other related fields.

Partnerships among zoos, botanical gardens, arboretums, natural history museums, geology museums, and aquariums are an obvious first step. It would mean breaking down the competitive barriers between these institutions and overcoming the distrust that assumes one specialist institution cannot do the job of another. All that requires is a shift in attitude. Nature writing, for example, now attracts a wide array of writers and a more avid group of readers, and zoos and botanical gardens, in partnership with local libraries, could play an active role by funding writers, hosting workshops, and holding public readings. Collaborative efforts with art museums would allow painters and sculptors to assist in carrying the messages of conservation and wise stewardship to new audiences. Working together, with shared programs and exhibits, zoos, botanical gardens, libraries, technology centers, and all other cultural, scientific, and natural history institutions could collectively engage in a public debate about new ways to look at Nature and about sound ecological practices in society, and devise many different ways to promote conservation. To save diversity, zoos themselves must become more diverse. The natural world can inspire painters as much as scientists, humanists as much as fundamentalists, scholars as much as dancers. Zoos have the potential to be the conductor of a new orchestration of conservation messages and activities that would enhance and even support the zoos' eminently worthwhile breeding programs.

DISTORTIONS AND EXCLUSIONS

The composition of zoo collections (an unsatisfactory word to describe all the living beings in a zoo's care) needs closer attention. The repetitive mix of charismatic megafauna—the lions, tigers, giraffes, elephants, zebras, bears, hippos, rhinos, and other large mammals found in almost every zoo—is hardly more than a hint at the richness of the world's diversity of animal species. It is a distorted presentation.

Zoos have traditionally perpetuated false images of animals, with bears in dank pits, lions in cramped cages, and monkeys behind bars. They now concentrate more on learning and entertainment, but their presentations can still be distorting. A continuing focus on a narrow band of wild animal species is conveying wrong perspectives to the current generation. Compare, for example, the composition of zoo collections with the diversity of animal species in the wild. Of the approximately 30 million species of animals on this planet, about 1,640 of them are mammals. The average American zoo collection contains 53 of these known mammalian species, a ratio of 1:31, or one type of mammal in the zoo for every thirty-one different species of mammals that exist in the wild. For birds, the ratio is less than one-third that figure, 1:98. It reduces even further for reptiles, with a ratio of 1:104. The deficiency is even more alarming in the realm of very small creatures. Amphibians are represented in the average American zoo at a ratio of only 1:2000. For invertebrates it drops to just one in several millions. Zoo collections reflect a reversed view of the natural world. More than 95 percent of all animals are each small enough to fit into a cupped hand but are unknown in zoos.

Admittedly, exhibiting these small life forms in zoos is problematic. They tend to have secretive lifestyles, are principally nocturnal, or dwell in places difficult to exhibit, like treetops or burrows. They often have very specific environmental needs, and because of their small size, they present a special challenge for close viewing. The devotion to displaying only live animals may have prevented zoos from exploring the possibilities of interpreting the behaviors of these tiny animals by using multimedia techniques. Some new ways to reveal their existence to zoo visitors and to tell the fascinating stories of their adaptations need to be found. Many of these very small animals have behaviors and lifestyles more interesting and illuminating than the traditional zoo species.

Aristotle made the point that "We ought not childishly neglect the study of the meaner animals because there is something wonderful in all of Nature. . . . We ought to investigate all sorts of animals because all of them will reveal something of Nature and something of beauty" (Krutch, 1961).

Invertebrates, especially, usually have more biological mass than any other species in the habitat and thus greater influence, with more vital and direct links to the functions of their ecosystems (Collins and Thomas, 1991). As E. O. Wilson has suggested, we need to better demonstrate that in many habitats it is often "the little things that run the world." These tiny life forms are not only invariably more critical to the survival of wild habitats, they are essential to our own survival. Wilson says, "It needs to be repeatedly stressed that invertebrates as a whole are even more important to the maintenance of ecosystems than are vertebrates. If invertebrates were to disappear, I doubt that the human species could last more than a few months" (Wilson, 1987). If zoo visitors understood this, they would have a quite different perspective and a better understanding of how our polluting and consumptive lifestyles, not just the hunters' and poachers' guns, cause such damage to wildlife. They would realize the terrible risks humans are taking by fouling the atmosphere, by spewing pesticides and other poisons all over the earth, and by burning and felling forests at such an alarming rate.

Humankind's ability to cause destruction gives us an equal responsibility. Learning to behave with more care, more awareness of our actions, and more affection for all things in Nature is an essential contract for humanity's future if we are to be anything but despoilers. Tom Lovejoy testified before Congress that the kind of rapid loss we are creating is a form of book burning and one of the greatest anti-intellectual acts of all time (Parsons, 1996).

The habitual focus of zoos on a small segment of the animal world raises the related question of their self-applied limitation to the field of zoology and the consequent exclusion of botany and geology. (The same problem exists with botanical gardens and their segregation from animals). It would be more useful for zoos to help us read the entire book of Nature, not just isolated chapters. The relationship between plants and their animal pollinators is but one example of a critical—and astonishing—story that is rather difficult to tell with only half of the cast of characters.

THE FORGOTTEN POLLINATORS

About 420 million years ago the first land plants developed from marine algae that had evolved a waxy skin and could avoid desiccation beyond the tidal zone. Simple and leafless, they were without roots, like mosses, but they could store water in their stems and thus hold their structures erect. In green and tangled masses, like miniature forests, they formed the habitat for the first animal colonists on the land. These were segmented creatures whose feathery gills evolved into breathing tubes, a trachea that allowed them to forage out of the water. With no competition these animals, which resembled today's millipedes, thrived. Some grew very large, up to six feet long. Inevitably, predators began to invade the mossy forests; they were the ancestors of today's venomous centipedes, scorpions, and spiders. The plants, too, were adapting to the new opportunities of life on land, becoming taller as they competed for more light.

As the first forests grew tall, they took much of their food supply in the form of spores and leaves up with them, out of reach of the animals on the floor. Multilegged creatures began climbing the trunks in pursuit of this nourishment. Others learned to take more permanent refuge in this new habitat, seeking cover from a new type of carnivorous invertebrate that had moved onto the land—the amphibians. Silverfish, bristletails, and springtails are descendants of the first insects that began climbing high in those primeval forests. Living silverfish have two small and rudimentary extensions of their chitinous shell on the back of their thorax. They serve as radiators, allowing blood circulating between these thin extrusions to warm more quickly in the sunshine, thereby allowing the animal to be more active as the energy-producing chemical reactions of their bodies proceed more rapidly. The extrusions look like rudimentary wings, and indeed insect wings today originate as small flaps on the back of the thorax, with veins that carry blood.

The first flying insects appeared about 300 million years ago, exploiting yet another unpopulated environment—the open skies. They diversified enormously. Cockroaches, grasshoppers, crickets, dragonflies, and many other insect forms began to fill the air. These new creatures played a critical role in a revolution that was underway among the plants.

Before the evolution of flowers, most plants distributed their male reproductive cells by broadcasting them to the wind within grains of pollen—an expensive and haphazard system. Insects visited these plants to feed upon their sugary sap and their pollen. Those plants that produced sweeter saps were especially attractive to insects and gained the advantage of losing smaller amounts of their less sweet and vitally important pollen. If pollen grains adhered to the body of an insect, they could be transferred to another plant of the same species, thus fertilizing it. This is a much more efficient system for transporting genetic material between plants. Some plants gradually adapted mechanisms that ensured more and more of this direct pollination.

Like humans and many other animals, insects are attracted to bright colors and to sweet scents. In some plants, nectar, which is a by-product of a plant's ability to photosynthesize, was produced in glands concentrated around the reproductive organs, optimizing the chance of an insect becoming smeared with pollen as it fed. Magnolias and waterlilies were the earliest and simplest flowers that we know of, first appearing about 100 million years ago. Their clusters of eggs are at the center of a group of stamens, which produce pollen. Modified leaves grew around this arrangement to advertise themselves to insects, eventually becoming what we recognize as blooms today. Beetles were originally voracious feeders of the pollen of cycads and pines but were among the first to become addicted to nectar and distribute pollens among the early flowering plants. Myriad other insects followed suit, then birds, and eventually some mammals. There are even day and night shifts. Brightly colored flowers attract day-flying animals while sweetly scented blooms advertise for the night flyers. The peculiarly shaped flowers of at least one tropical velvet-bean vine reflect ultrasound pulses in a focused beam from considerable distances, thus guiding nectar-eating small bats as clearly as radio signals guide airplane pilots. Some flowers, such as fuchsia and honeysuckle, have developed a strategy to ensure that individual flowers are visited often by the pollinators, by producing nectar only in small amounts at intervals. Some plants provide landing pads for insects, and the trail of speckles on a foxglove act like landing lights, guiding the insect directly to the nectar, and the location of the flower's sexual organs. There are even cooperative blossoming strategies between some plants: as one flowering cycle wanes, another species' cycle begins to peak. When different species bloom at the same time, they have methods of depositing pollen on different parts of a pollinator's body.

One problem with having eggs and pollen in the same flower is the possibility of self-pollination and the subsequent loss of genetic dissemination. Thus, eggs and pollen mature at different times. Magnolia stamens will not produce pollen until after insects bringing pollen from another magnolia have fertilized the eggs in the flower. It is also important, however, that pollen is available when the eggs are ready, which is why plants of one species all generally bloom together.

Today, the majority of the world's quarter million species of flowering plants depend upon animals, sometimes in exquisitely specific partnerships, for pollination. Not even the basics of this phenomenon seem to be known by the general public. In a study by Nancy Pratt at the National Zoo in 1995, among visitors with educational ranges from grade school through professional degrees, fewer than one-third knew that a plant's flower was for reproduction, more than 40 percent thought pollen nothing but a nuisance, and although more than 75 percent knew that bees make honey, about half of the people gave a negative response to the word "bee."

Complex interdependencies that have evolved between plants and pollinators over millions of years have suddenly become extremely vulnerable, mainly because of pesticides and loss of wild habitat. It is estimated that more pesticides are now sprayed over suburban gardens than were ever used by agriculture in the days when Rachel Carson so eloquently warned us about the risk of a silent spring. It remains a frightening possibility that for the sake of blemish-free apples and unnibbled begonias, we are threatening to undo the complex interactions that began the extraordinary cooperative evolution of plants and pollinators hundreds of millions of years ago. We menace our own future as well.

Without the critical pollinating interactions between certain plants and animals, say bee expert Steve Buchmann and ethnobotanist Gary Nabhan, we could lose the seeds, fruits, and plants that make up one in every three bites of our daily meals. These are food items that comprise nearly all of our dietary diversity and would leave us with a diet essentially limited to meat and grains; no tomatoes, no berries of any kind, no peppers, almonds, squash, apples, oranges, bananas, or any other fruit, not even any chocolate (Buchmann and Nabhan, 1996). Having unlimited access to uniformly ripened fruits and vegetables in our supermarkets has shielded us from recognizing how much we rely upon these wild and vital processes, many of which are now in jeopardy. A survey of 1600 Australian schoolchildren (Kondonin Group, 1997), for example, found that 18 percent believed people in cities did not need farmers and only 55 percent believed that the food they ate for dinner came from a farm. What are the chances that they understand the extended impact of the death of the bees they just swatted or the pesticide they sprayed into the air? And if zoos don't tell these stories, who will?

NATURE IN THE BACKYARD

People now speed along the Information Highway rather than walking country lanes. Children browse the World Wide Web rather than watching a spider spin. They wander the Internet more than the local woodlands. Instead of exploring wild streams, they splash in concrete gutters. They are more exposed to amplified music than to birdsong and spend more hours in malls than in meadows.

Information about the natural world once held secure by older generations is no longer in our collective memory. It seems perverse that children routinely assemble more facts about dolphins or tigers for a homework project than their grandparents ever accumulated in a lifetime, but that their knowledge of local wildlife is minimal. Zoos face a responsibility and a challenge of utmost importance in this regard. A greater focus on regional conservation issues would be a positive move and direct action would be even more valuable.

Visitors to San Diego Zoo can hear messages about the threat of tiger extinction and nod their head in concern, then drive north to the San Diego Wild Animal Park and hear messages about the depredation of elephants by poaching and comfort themselves by agreeing that they will never purchase ivory. The road they will have traveled between these sister zoos will have carried them through a region in which virtually every square inch of native chaparral habitat has been destroyed. It has been replaced by suburban sprawl and monoculture farms. Southern California contains one-fourth of all plant species known in the United States, half of them found nowhere else in the world, and it is one of the most endangered ecosystems on Earth. Yet there will be no hint from these two Southern California zoos revealing this problem. It is too uncomfortably close to home. This dilemma is not exclusive to California, of course. It is, however, frustrating that it is ignored by a zoological society such as San Diego's, willing and able to pay one million dollars every year to the Chinese government, which they say will be used for wild habitat projects in China, in exchange for the loan of a pair of giant pandas. Exhibiting a pair of giant pandas significantly increases gift shop sales and profits. Trying to make zoo visitors realize that *local* wildlife habitats are being wiped out will result in mail to the zoo director consisting of difficult if not vituperative letters from local developers. Which, then, would be the correct course of action for a zoo that, like San Diego's, claims to be "dedicated to preservation of our planet's precious ecosystems and wildlife": educating the public about local destruction of wildlife habitat or placating developers with silence and entertaining zoo visitors with million-dollar exhibition animals?

Rather than paying attention to local and regional conservation, there has lately been a mania among zoos for building tropical rain forest exhibits, usually of massive size and cost. It mirrors a trend in society: general discussions about saving the wild seem to have become synonymous with saving tropical rain forests. Undeniably, these forests are suffering appalling damage, but this fixation by zoos comes to be seen as a concern remote from their visitors' daily lives. More comprehensive presentations by zoos would help people develop views of both global and local significance.

It has been traditional for zoos to engage their principal efforts on exotic species, ignoring regional problems. Australian zoos, unable to import several exotic species, have been forced into the lead in correcting this problem. European zoos could benefit from such an approach, for there is likely to be as much pressure on the remaining small pieces of wildlife habitat in Europe as in any other region on Earth. Zoos in developing countries could make a useful contribution by bringing attention to locally rare species.

In North America, the world's most diversified temperate hardwood forests are under severe threat. This brings the potential for linked extinctions of many animal species. Large numbers of tree species in these forests are dying from pollution. Virtually every American zoo presents visitors with masses of statistics about depletion of tropical rain forests, but few if any mention the uncertain future of their own nation's hardwood forests.

To casual observers these forests might still look healthy, but environmental journalist Charles Little (1995) tells a sobering account of declining forest biomass across the continent. This is not only an ecological problem but also a political issue, because the way to regain healthy, biologically diverse forests in America is to eliminate the acid rain, heavy metal contaminants, and smog that are polluting them and to abandon the policies of clear-cutting and fire suppression. These processes, however, have also created wealth and power and have sustained jobs. People must know the full range of their options if they are to make wise choices on such issues and must be very well informed if they are to evaluate their personal and immediate wants against the long-term sustenance of ecosystems.

Landscape architect Laurie Olin (1996) describes driving past an entire mountainside in the Tuscarora Mountains of Pennsylvania that had been clear-cut, leaving miles of desolation. He notes how this sort of despoliation has become so pervasive "as to be hardly noticeable to most Americans and is often perfectly legal, part of the desperate economic plans of state and federal agencies." Zoos must do all they can, with energy, imagination, and enthusiasm to show people that the region in which they live is precious and in need of intelligent stewardship.

In support of this, zoos could advertise the conservation practices they are pursuing that other local businesses could adopt. Is the zoo offering employees incentives for carpooling or using mass transportation? Is it actively engaged in recycling? Does it use only air conditioners that don't emit chlorofluorocarbons? Is it dedicated to using minimum amounts of electricity? Do its construction projects follow the principles of green architecture? Does it have ecologically sound purchasing policies for selecting materials and products? These practices, to borrow London Zoo's current advertising slogan, would be "Conservation in Action."

BROADENING THE VIEW

In the past, all attention in zoos, by keepers and by visitors, has been upon specific, individual animals. Oftentimes some particular chimpanzee or elephant or bear (it is almost always a big mammal) has been packaged as some sort of national pet. It is a habit that mirrors the Western preference for seeing the tree rather than the forest, a reductionist perspective with an obsession for icons that represent a complex whole. This simplistic view could have profound effects upon conservation efforts. The public finds it easier to give money to help Elsie the zoo elephant than to help save elephant habitat.

Whether we like it or not, human intervention in the lives of wild animals must inevitably increase in decades to come, as ever more wild habitat is lost, mutilated, or fragmented. These actions will cause new difficulties since conservationists will focus on protecting species populations and the processes of natural systems, while the public and the media will want to place their attention on individual animals. There will be conflicts, but perhaps zoos can play a role in helping engage their visitors in questioning what we really value in Nature. Should we, for example, kill feral rabbits in Australia? Few would deny that, for they are so destructive. Would there be the same support for eradicating feral cats? What about killing or removing feral goats that are destroying native habitats on islands? Or feral horses? Where is the line that separates concern for an ecosystem from concern for individual animals?

It is not unusual to find ourselves in a dilemma evaluating human life against an animal's. How legitimate is it to develop new vaccines that could save the lives of human children if it requires testing monkeys to death? Can we ethically torture animals to produce tear-free shampoos for our babies? And what effects do these assaults on animals have upon our very humanity? Our relationships and attitudes to animals and to the wild are hopelessly muddled. When a young woman jogger was killed in California's Sierra Nevada foothills by a mountain lion in April 1994, the lioness was hunted down and shot. The woman left two motherless children, the lioness an orphaned cub. A public appeal raised nine thousand dollars for the children and twenty-one thousand dollars for the cub (Cronon, 1995a). This is a perplexing statement on human nature and on human attitudes about Nature.

If you see your cat stalking a wild bird, you have a moral obligation to intervene and scare the cat away. If you see a cheetah stalking a wild gazelle, your moral right is abrogated. Why? Because one situation is an artificial context—humans have spread the domestic cat around the world—but the other scene is part of Nature, operating on a system that we have no duties to control. What, though, of a situation in a national park, where wild animals have been restricted to much smaller areas than their former range? When a bison fell through the ice into the Yellowstone River, it was allowed to drown, because the park authorities made the difficult decision to sacrifice the individual for the sake of allowing a natural system to operate. Some would say this was inhumane. Is paving fertile valley habitats to build a shopping mall essentially human, or inhuman? If we are to accept the responsibilities of managing the earth, which is a position we have stumbled our way into, we must not only discover how to find answers to such questions but also learn to ask such questions in the first place.

In zoos, these dilemmas have become especially acute. Plans were proposed a few years ago to move a gorilla from Cleveland Zoo to the Bronx Zoo, to improve his chances of producing offspring and thus contributing his genes to the shrinking genetic pool. The proposal generated violent debate in the media, plus litigation from an animal rights organization that considered the move would be sufficiently traumatic to the gorilla to outweigh any scientific or conservation value. One side was arguing for the good of an individual gorilla, one for the good of the gorilla species. (The gorilla was eventually relocated and became a father at the Bronx Zoo.)

There is potential for many such conflicts in the future. Managing a species means managing individuals of that species. More and more, management of the wild will interact with and impinge upon people and what they perceive as their rights and needs. What scientists call wilderness will be, to others, a homeland in which they have evolved a way of life of hunting, farming, tree felling, and grass burning. No wonder that the Bronx Zoo's worldwide conservation program, for example, is no longer restricted to involvement of biologists, but now also includes lawyers, economists, sociologists, and anthropologists. Conservationists, warns Conway (1995b), are going to have to become supreme pragmatists in the future. Thus the New York Wildlife Conservation Society, while establishing new reserves in Bolivia, Brazil, and Madagascar, has also encouraged hunting in Zambia, avoided condemning a hydroelectric project in Laos, set up eco-tourism facilities inside wildlife parks, and helped people in Patagonia farm inside wilderness areas. In each instance they decided that such actions would be more advantageous to conservation in the long run, even though they knew that there would be detrimental impacts upon individual animals in the short term.

Admittedly, in shifting our sympathy from individual wild animals to a concern for preserving biodiversity and ecosystems, there is a risk of stepping into methods of evaluation that are nothing but coldly scientific. It is a quandary that must be approached with great caution. Whether we can learn to love Nature as much as we empathize with Bambi, need not, however, be exclusionary: it is not necessarily a case of either one or the other. Many people with a general love for wild species have formed their compassion based on experience with one individual animal. In this way, zoo animals can become powerful ambassadors for their wild species. (And as such must be treated with the respect and care that ambassadors are due.)

Even so, the truth is that Nature will not work by our standards or our ethics. If we have inherited a world that is so disjointed by human action that it now requires our intervention to heal and repair it, then we will have to learn to accept that we must try to manage it on the basis of Nature's systems and not our standards. But we have much to learn to become better managers. Simply setting aside wild lands, for example, is not always sufficient, notes Conway (1995b). The Everglades National Park was established in 1947 with the goal of protecting "the finest assemblage of large wading birds on the continent." By 1989 this finest assemblage had declined by 90 percent. A study of seven of the largest national parks in western North America, in 1987, revealed that twenty-seven species of mammals had become extinct in these protected reserves. In 1980 the Indonesian government, assisted by the World Wildlife Fund and the International Union for Conservation of Nature, published *The Javan Tiger and the Meru-Bitiri Reserve: A Plan for Management.* It overruled any efforts at captive propagation, relying solely upon habitat protection. Today the Javan tiger is extinct.

As one of the potential key players in wildlife conservation, zoos need to learn to broaden their focus beyond animals and to encourage a wider vision. Our world has an urgent need for natural history institutions that present Nature, not only its species, and that interpret Nature, not just a narrow spectrum of its animals. Our ultimate aim must be to conserve Nature, in all its complexity, diversity, and fragile beauty.

There is movement in this direction. More zoo professionals are recognizing that habitat conservation, and not only species preservation, is critical and are becoming involved in field conservation. Increasingly, exhibits in zoos are being named as presentations of a habitat rather than of an animal species. Greater efforts are being made to create total habitat experiences for visitors rather than simply putting animals on exhibit. More zoo education staff are trying to develop holistic interpretations of the natural world. They should not only be given maximum encouragement for this approach but also have a stronger role in forming zoo policies and strategies. Too often educational staff have only middle-management status, usually even lower than the public relations staff, in spite of their critical role in forming the future of zoos.

Michael Robinson (1988), director of the National Zoo in Washington, D.C., has for more than a decade been promoting the concept of bioparks. He advocates broadening the definition and the content of zoos to focus on biology, to show the connections and interdependencies between plants and animals, and "to educate for a bioliterate citizenry." He notes that we would more fully appreciate how marvelous an elephant or an ant is if we understood the animal's skeletal mechanics, blood circulation, evolutionary history, and relationships with other plants and animals, and if we could thereby recognize the ties that bind us and all living things.

Zoos need to move forward with such thinking more quickly and with greater urgency and daring. The efforts of all of our natural history institutions in total are failing to teach even the basics about our planet and how it works and what we must do to protect it from ourselves. In particular, democracy faces an urgent dilemma in its need for people to vote for politicians who care for the environment and who will make decisions beyond a pork-barrel mentality. It is not an easy strategy. Most politicians are attracted to the quick fix, to something that shows results during their term of office. Conservation efforts, by contrast, require the sort of care that considers the effects of today's actions upon unborn generations. When Margaret Thatcher told British Conservative Party members at their 1982 conference in Scotland, with reference to the Falklands War, "It is exciting to have a *real* crisis on your hands, when you have spent half your political life dealing with humdrum issues like the environment," she was revealing a basic problem in both her philosophies and her profession. Recently a Republican congressman from California introduced legislation against the Endangered Species Act so that construction of flood levees in his state would not be delayed "for bugs and rodents." We have a scale and a pattern of ecological illiteracy in our modern society that is terrifying. Yet millions and millions of people visit our zoos each year, many with open minds, craving contact with and clearer understanding of that other world of Nature.

CHANGE AND EFFECT

It is time for zoos to reexamine their philosophies. People no longer need to visit a zoo to see what a camel or a leopard looks like so much as to gain a better understanding of the dynamic systems of Nature and the interconnections within ecosystems, and most especially how to help conserve biological diversity on the planet. Zoos need to make the concept of biodiversity not just intelligible but *wonderful* to their audiences. It must be seen not just as fascinating, but as absolutely critical to the continuing health and well-being of people. It must come to be seen as something worth making sacrifices for, something sufficiently valid to warrant changes in our lifestyle.

When the thirteen American colonies set out to win independence from Britain, their decision to boycott British goods required severe and immediate hardships. They made the choice to do this because of their faith in an ideal and for the eventual safety and well-being of their descendants. Today we face a similar problem with regard to protecting the health and viability of our planet as a good and productive home for our descendants. Thus, developing alternative energy supplies, recycling materials, investing with social responsibility, giving sanctity to wilderness, minimizing population growth, avoiding development on critical habitats or other productive land, and learning ways to save water can all be seen as revolutionary values similar to those forged more than two hundred years ago on the eastern seaboard of North America.

The environmental ethics of wise stewardship of natural resources, of balancing spiritual and material wealth, of ensuring that the next generation inherits a world as rich and healthy as the one that welcomed previous generations represent a growing concern about the interdependence of human culture and natural systems.

Zoos have little experience proselytizing, marketing, or teaching values, so zoo educators will need to hone new skills to increase their effectiveness. But *more* conservation education is not all that is needed. Wilderness advocate Aldo Leopold (1966) asked whether it is "certain that only the *volume* of education needs stepping up? Is something lacking in the content as well?" In the best zoos, visitors see animals in replicated habitats and may gain some respect, leaving better informed. In other zoos they may find animals in misery or see an animal show for which the primary goal is to generate excitement and laughter. Television news, with rare exceptions, has been reduced to infotainment, and many zoos unblushingly and earnestly seek to reduce education to edutainment. Moreover, what passes for education in too many instances is a mere stream of simple facts: "Did you know that giraffes have seven vertebrae in their neck, just like people?" What we need to *know,* instead, is our dependence upon the natural riches that sustain us, emotionally, spiritually, and physically. The messages of interdependency, however, are not being relayed; the lessons of interconnections are ignored. Zoos must take on the challenge of persuading people that environmentalism, far from being the dirty word that some claim, is worth fighting for. Without an informed citizenry willing to support conservation, people will not only continue felling the ancient trees but will eventually saw off the very limb on which humankind is sitting—not that humans are likely to wipe themselves out of existence, but a very nasty fall is likely. Unless we take stronger action to conserve wildlife habitats, we will continue to decline into a morass, a world devoid of much of its current complexity.

Humans are too widely distributed to join the millions of other species doomed to extinction by our destructive habits. Many other species will also survive and even proliferate because of human activities. Aggressive and opportunistic generalists that flourish in human-degraded environments will take advantage of our destruction of wilderness areas and will crowd out many specialized life forms. The earth will lose thousands of diverse life forms and have them replaced with rats, cockroaches, pigeons, coyotes, ragweed, fire ants, mosquitoes, tamarisk trees, kudzu, zebra mussels, cowbirds, sparrows, flies, silverfish, feral cats and dogs, thistles, spurge, gray squirrels, crows, and house mice. These will become the widespread and populous species that will go forth with us and multiply around the world, filling the vacant niches that humans have created (Quammen, 1998). A complex and diverse biological heritage will be supplanted by a degraded population of weed species, with enormous economic, environmental, and emotional impacts on humans.

Because modern life has removed direct contact with the diversity of the natural world, we may not recognize the dangers of a pauperized environment. As Nabhan and St. Antoine point out in their 1993 study of the loss of flora and fauna, humans rely almost entirely upon a mere handful of plants and animals for food: 50 percent of the world's caloric and protein intake, for example, comes from grains. Just twelve crops are the basis for 75 percent of the world's human food supply (Groombridge and Jenkins, 2000). We operate on a disturbingly narrow biological base. This has led to a false sense of security and a misunderstanding of the biological safety net that lies beneath the technological tightrope along which we are shuffling our way into the future. Biodiversity is not an expendable commodity we can manage without. It is not a mere luxury. If we lose any of the endangered zoo species it will be heartbreaking. But if we do not maintain biodiversity in our wild places, the results will be far more profound than a sense of deep sadness.

A shift away from their concentration on exhibiting only animal species—especially in the tidy arrangements of taxonomic groupings—and a move toward revealing the stories of Nature, of ecosystems and biological communities, will bring enormous challenges for zoo professionals. But solving very difficult challenges brings the greatest satisfaction. Best of all, and of the most enduring value in this instance, it would bring untold benefits to those young people who, maybe even more than their parents' and grandparents' generations, desperately need to understand the complexity and fragility of the world of Nature and their place in it.

CHAPTER EIGHT

WHICH WAY THE FUTURE?

Few activities are more seductive, or more futile, than predicting the future. Time is always making a fool of the pundits, yet they persist. Perhaps people need to paint the picture they hope the world will become. In the 1920s and 1930s, when modernism was starting to surge, an eruption of books and movies enthusiastically prophesied the glorious wonders ahead. Atomic and solar-generated electrical power was to beam us into a brave new world, providing unlimited clean power for everything from rocket ships and gyroscopic automobiles to unlimited good health.

Trying to divine changes in public attitudes is equally problematic, complicated by the fact that people in different cultures think in different ways and have different standards. As biologist and social historian Alex Comfort (1966) points out, people in one culture "would admire what the other despised, and be ashamed of what the other would be proud to do." One manifestation of this is the world's widely varying attitudes about the treatment of animals. Large-scale killing of wild animals for "aphrodisiacs" and medicines is openly practiced in some places, scorned as horrific in others. People in different countries, as much as those in distant days, can hold widely discrepant beliefs from one's own time and place.

Distinctions in zoo standards around the world are fairly clearly defined within a given culture, but the interwoven tapestry of factors that supports

A pigtailed macaque at the Johore Bahru Zoo, Malaysia, 1988. Uncounted masses of animals suffer in disgusting and cruel conditions that prevail in zoos across Asia. (Author photo.)

such standards remains a mystery. Some cultures routinely subject animals to abuse, while others, particularly in northern Europe, tend to have more humanitarian attitudes or at least to have promulgated animal welfare legislation. Not that any of this is consistent; consider fox hunting in England, kangaroo shooting in Australia, and activities at primate research centers in the United States. Grouped patterns exist, but they do not seem to be determined by any specific economic, religious, or other obvious causes, which is puzzling. Zoos of generally poor and even barbaric standards, for example, are common throughout almost all of Eurasia, from the nations of Eastern Europe all across the continent to Taiwan, as well as in all countries of Latin culture. Yet there are extreme variations of wealth (Japan versus Pakistan, for example), of beliefs (from Hinduism to Communism), and of social histories (Greece compared to Uruguay) within these regions. Some regions seem to specialize in vile zoos. An abundance of them proliferate throughout

South America and across the whole of Eurasia, and some of the world's worst are the products of communist governments. Mix this with cultural differences, and the strange results are zoos throughout China that display cats and dogs in filthy little cages. The conditions at some Chinese zoos defy description. The filth and stench at Haikou People's Park Zoo, on Hainan Island, is indescribable. Kunming Zoo, in Yunnan Province, includes a pond from which fish are caught on hooks all day and repeatedly put back for catching again. At the Siu Ling Park, Yanzhou, visitors to a dark and filthy building are given a stick to jab the monkeys. In many of China's zoos, visitors routinely and contemptuously spit upon the animals.

People seem to enjoy having their photographs taken with wild animals. In some countries many creatures suffer for this. At Xili Lake Zoo, in Shenzhen, you can sit next to a drugged, declawed, defanged tiger while he is struck on the mouth to make him snarl for the camera. A bear is forced to stand and bare his teeth by a string that runs over a pulley and is attached to a ring in his upper lip. Similar animal photography opportunities abound not only in China but also in southern Europe and all the Latin American countries.

Having noted the characteristically abysmal zoo standards in these widely different areas of the world, however, one must also name the exceptions. There are progressive thinkers at the zoos in Barcelona and Lisbon, for example. The Night Safari, the Zoological Gardens, and the Jurong BirdPark in Singapore, as well as Zoo-Ave in Costa Rica, the Belize Zoo, and the zoo in Tuxtla Guitiérrez, Mexico, are as good or better than anything to be found anywhere else in the world.

Nor is the unpleasantness of poor zoo standards absent from Western Europe and North America, where the general standards are high. Horrible zoos can be found in those regions. It is very difficult to use the word "zoo" without apologies for these wide variations. The U.S. Department of Agriculture issues permits to over two thousand zoos. Only 186 of these are accredited by the American Zoo Association. Melbourne Zoo, Florida, is one of the nastiest; Melbourne Zoo, Australia, is one of the best of zoos. That they share the same name is ludicrous. The Woodland Deer Park Zoo in California bears no resemblance to the Woodland Park Zoo in Washington. Terry Maple, director of Zoo Atlanta, refers to nonaccredited zoos simply as "impostors." It is as good a term as any.

SEEKING COMMON GROUND

Surely, one day, accredited zoos will work toward some form of legislation to ensure global standards and abolition of the substandard impostors. At present this thankless task is left to animal welfare societies: licensed and recognized zoos have to date largely refused to enter the fray.

In the very best zoos, wild animals can be seen as ambassadors for the survival of their species in the wild. In the worst zoos, they generate nothing but negative reactions, bringing only tears and anger, or, to the insensitive, confirmation of their own malice. The frustration of an interchangeable title for places of such widely varying standards and purpose is more than merely academic. For one important example, it confuses the argument about standards to the point that emotionalism takes over rational thought and becomes absurd. The passion of those who believe that zoos should be abolished blinds them to an acceptance of the fact that zoos are firmly established institutions around the world and the fact that they must work from that reality. The abolition of zoos is extremely unlikely. They will not be wished away.

Animal lovers who work in zoos, and animal lovers who work against zoos, would serve wildlife better if they abandoned their cold war, searched for their points of common ground, and engaged their combined energies and expertise to develop higher and universally accepted standards. There have been encouraging signs that the silent hostility between animal welfare agencies and zoos is thawing. The Georgia Institute of Technology hosted a workshop in 1992 (the proceedings, *Ethics on the Ark,* were published by the Smithsonian Institution in 1995) in which zoo professionals, animal welfare activists, conservation biologists, and philosophers engaged in a sometimes passionate discussion about the future of zoos and the treatment of animals in captivity, and the question of whether it is the individual animal, the species, or the ecosystem that should be the most important focus in any wildlife conservation efforts. In part the participants only agreed to differ. Some advocated an increase in captive breeding programs. Others called for zoos to focus on community education programs and to pay more attention to protecting local habitats than exotic species. Still others contended that zoos had no moral justification for existence, whatever scientific or educational purpose they pursued. Yet there did emerge a consensus that the natural areas of the world are fragmenting and shrinking, and that national parks and wildlife preserves must inevitably take on some of the character of zoos. Thus the questions confronting zoos, of how to ethically proceed into the future, become questions for the planet and the world of Nature.

More recently, a conference sponsored and organized by the Bar Association of the City of New York engaged a wide range of professionals from the world of zoos, animal-welfare societies, philosophy departments, and nongovernmental agencies in a debate moderated by lawyers with personal interests in wildlife issues. As recently as ten years ago, such a gathering would probably not have been possible and certainly not productive. There was too much suspicion on each side. It is a mark of considerable progress that this 1998 symposium even took place. Most encouraging were the promises of AZA officials to work with outside parties to lobby the U.S. Department of Agriculture to increase the level of fees paid for permits to operate zoos. At present these fees are ridiculously low and of no consequence to roadside zoo and nonaccredited operators. The same level of agreement was forged to try and persuade the USDA to more aggressively prosecute instances of neglect and abuse under the Animal Welfare Act. There were open discussions and active debates, too, on issues such as the unsatisfactory conditions under which many zoo animals are held in off-exhibit holding areas, and on the need for zoos to recognize a cradle-to-grave responsibility for creatures born in their care.

While the 1992 workshop was in process in Atlanta, a more immediately tangible and specific problem was being tackled in Seattle. The Progressive Animal Welfare Society (PAWS), guided by Mitchell Fox, launched a campaign in the early 1990s to find a better home for an adult male gorilla named Ivan, displayed in solitary confinement in a cage in a department store in Tacoma, Washington. Working with great energy, fortitude, common sense, and some lawyers, Fox eventually won this long battle, but only with the strategic assistance of David Towne and Terry Maple, directors of the zoos in Seattle and Atlanta, respectively. After years of effort Ivan is today successfully integrated as a new father into a family of gorillas in the world's largest naturalistic exhibition habitat for the species, at Zoo Atlanta, and is living testament to the progress that can be made when people of shared commitment come together from different perspectives.

Tackling projects of this complexity is extraordinarily difficult—physically, legally, financially, and especially emotionally. There surely cannot be any members of the American Zoo Association who would approve keeping Ivan in the conditions that he had to endure for so many years, yet their professional silence was deafening and even gave tacit support. The same conditions exist in many roadside zoos, and it is to the zoo community's shame, and, ultimately, detriment, that it will not risk tackling these obscenities. The problem of dealing with the worst excesses of captive animals kept in intolerable conditions has been taken on by Virginia McKenna and her son, Will Travers. Yet their worthy efforts have earned them only hostility from many zoo directors. Their organization, the Born Free Foundation, established in England in 1984, has many active divisions, including Zoo Check, which works to investigate, expose, and rescue animals from slum conditions in zoos around the world and to prosecute the owners. For more than a decade the foundation has also been actively campaigning against wildlife poaching in Africa. Both of these programs are natural partners for zoos dedicated to ethical standards, and there is an element of ridiculousness in the zoo profession's unwillingness to provide assistance. The Humane Society of the United States contrasts the management of accredited zoos with the "deprived, decrepit conditions" in many other so-called zoos. They rightly complain, though, of the established zoo profession's "conspiracy of silence that leads directly to acceptance of lowest-common-denominator conditions" (Grandy, 1992).

THE ELECTRONIC ZOO

The Born Free Foundation has developed plans for a worldlife center, to be built in Leicester, England, naming it TechnEcoZoo and promoting the idea as "The 21st Century's non-animal answer to the traditional zoo." The center will interpret the formation of life on Earth, the current environmental crisis, and the history of human relationships with wildlife, using only a variety of technologies and avoiding any live-animal displays. Work has not yet begun on the site for this project but another sort of electronic zoo opened in 2000 in Bristol, England, called Wildscreen. It is the brainchild of Christopher Parsons, who was head of the BBC's natural history unit for many years.

The genesis for the idea has several starting points. Parsons has been aware for longer than most that well-designed, thoughtful, and inspiring stories about Nature can have extraordinary consequences. Consider the impact of David Attenborough's television series *Life on Earth* and *The Secret Life of Plants.* Parsons produced the *Life on Earth* venture for BBC Television and says that he would be a rich man, "if I had £100 for every graduate zoologist who has come through my door and mentioned that [this series] had set him or her off on their career."

After retiring from the BBC in 1988, Parsons was contacted by Imax Corporation, the Canadian company that pioneered large-format film systems, to discover whether the medium could be used to create natural history films for museums and zoos. He soon realized that "large-format film technology could not have been less suited to producing natural history films." The cameras for this medium had to use wide-angle lenses; were large, heavy, and noisy; and consumed film at prodigious rates. By contrast, the best natural history films have been made by individuals working alone or in small groups, using lightweight cameras with zoom lenses. But Parsons also knew that the immersion effect of IMAX films offered something extra. Television had proved itself superb in conveying details of Nature and animal behavior to mass audiences in an intimate way. Large-format film, conversely, could convey grandeur and spectacle beyond the capacity of television. New camera designs, he conjectured, and careful selection of subjects, could allow large-format natural history films to be produced. This was the first building block in the foundations of the idea for Wildscreen.

Very soon after this, 3D large-format film arrived on the scene. The first underwater film in this new medium was a natural history subject, *Into the Deep,* directed by Howard Hall. It opened first in the heart of Manhattan, at Sony's new 3D IMAX Theater, and was amazingly successful. The visual and emotional effects of this new technology are inspirational. Hall says that when he went to the theater to watch his film, it was, "apart from the fact that I was not wearing all my underwater gear, exactly as I saw it when I filmed it." The potential for Wildscreen had taken a quantum leap.

The second component of the project had been in Parson's mind since the early 1980s, when it had occurred to him that a huge audio-visual record of the world's living organisms was being compiled by various people and that it had enormously useful potential, if properly coordinated, as an educational resource of incalculable value. His idea was ahead of its time, but with the arrival of digital technology and the Internet, he now had the means to make this vast resource available all over the world. Hence, the inclusion of ARKive, a digital library of images and recordings of endangered species, in the project, as a global biodiversity resource network.

The third component is a themed exhibition that seamlessly combines the live presentation of animals and plants with advanced audio-visual techniques. While working at the BBC, Parsons had often been approached by zoos interested in using film footage in various types of audio-visual enhancement of their exhibits, and in 1982 he noticed a great opportunity in a temporary display that the BBC had produced to celebrate the twenty-fifth anniversary of their natural history unit. The exhibit showed equipment for specialized techniques in filming very small life forms. Focused on emerging mosquito larvae, it attracted unexpectedly strong attention. Parsons concluded that the appeal for people was that it not only allowed visual entrance into a tiny and otherwise invisible world, but it was also live action.

Later, when he first saw high definition television (HDTV) and was impressed by the clarity, size, and steadiness of its images, he "had the impression of looking at the natural world through a kind of magic window." He experimented with these techniques and discovered that when HDTV live pictures were extended to include three large, contiguous screens that almost encircled the viewers the effect was even more remarkable, similar to being in a wildlife observatory. When Parsons launched his Magic Windows concept in a casual-gathering atmosphere with cocktails and snacks, he found that people were ignoring the opportunity to chat, drink, and eat, but were instead watching transmission of birds on mud flats—oystercatchers at Morecambe Bay in the north of England. The birds' activities were not particularly riveting material, but the viewers seemed mesmerized by what they were seeing. Parsons conjectured that the secrets, once more, were the film's perfect clarity and the fact that it was live. As he points out, one can spend half an hour or more watching elephants at an African waterhole, not perhaps actually seeing much, but absorbing a visual experience that is memorable and magical. If it had been an edited film, he realized, people's attention might soon have wandered. But because of its slowly unfolding real-time action, its urgency and immediacy were compelling, almost spell-binding. In addition, there appeared to be something more deep seated at work. Observations by curator Dale Marcellini at the National Zoo's reptile house, which presents animals in very traditional zoo fashion, showed that visitors spend an average of only eight seconds at each exhibit, which is typical of visitors at many zoos. Parson's HDTV transmissions were holding people in rapt attention for twenty minutes and longer. The essential difference, the key component to the attraction, was the perfect authenticity of the scenes that people were watching. The National Zoo's reptile displays, like most zoo displays, are clearly artificial and decidedly unattractive, too. They have none of the inherent beauty of minute detail found in scenes from Nature. The HDTV transmissions of an authentic wild place were so convincing and immediate that people felt involved and wanted to watch and learn. If zoos could replicate such detail and authenticity, they, too, would captivate visitors for hours.

Parsons gradually became ever more aware of the desirability of presenting the public with information about not just animals, but the entire world of Nature: a notion far removed from the traditional focus of zoos. He thus evolved the concept of Wildscreen: an attraction that was neither zoo, botanical garden, nor museum, but which embraced elements from all these institutions. And because much of the natural world's fauna is very small, he would show minuscule live animals as well as dramatic and spectacular natural scenes, all enhanced by film clips of relevant behaviors on demand (Parsons, 1996).

Large-format cinema, HDTV, the Internet, and archival film footage, as well as natural history museums, aquariums, botanical gardens, and zoos all provide possibilities for convincing people to perceive Nature in a more holistic way. All these techniques come together for Wildscreen.

A BROADER VIEW AND CLOSER

The visitor's experience of Wildscreen begins with brief but dazzling three-dimensional multimedia presentations: one showing the planet's biodiversity with wonderful images of differing shapes, colors, sizes, lifestyles, and habitats and another the time scale of life on Earth and the constantly changing patterns of land masses and climates that accompanied and directed its evolution. Wildscreen then introduces exhibits that explain the possible origins of life. A mix of displays demonstrates the essential differences between the five kingdoms of life—bacteria, protists, fungi, plants, and animals. (Protists formed as a radically new type of cell, about 2 billion years ago, when respiring bacteria, after hundreds of millions of years, had raised oxygen levels in the planet's atmosphere to more than 20 percent. Protists have cell parts, known as mitochondria, that consume oxygen. The degree of separation between the new protists and the aerobic bacteria is, says Amherst's influential biologist Lynn Margulis (1986), "the most dramatic in all biology." Plants, animals, fungi, and protists are all based on the nuclear cell design, reflecting a common heritage. They differ radically from the bacterial kingdom.)

In the first main exhibition hall, visitors see many kinds of planktonic life forms in 3D electronic displays and live displays of marine invertebrates before discovering the first animals with backbones and skulls—fish. An adjacent botanical house is arranged with a succession of plants, from the first mosses and liverworts to the recently evolved flowering plants, illustrating the story of land invasion by plants. Interpretive exhibits tell the stories of the extraordinary adaptations required by plants to transfer from life in water to life in air, such as solving structural engineering problems and finding new methods of reproduction.

A separate exhibition space follows, dealing with the story of the next wave of invasion by animals from the seas and how they evolved new methods of breathing and locomotion. A large group of exhibits shows the diverse form and behavior of insects, which account for more than two-thirds of all animal life. A large section is devoted to arachnids, including scorpions, spiders, and mites. Only a comparatively small amount of space is devoted to amphibians, reptiles, birds, and mammals, reflecting their very low species numbers compared to all other animals.

An intriguing assemblage of exhibits presents the world of fungi, correcting the common misapprehension that they are plants and showing their huge significance in the world's ecological systems. Without fungi more than three-quarters of the world's plants, and especially trees, could not exist, which in turn means that humans could not either. The story of fungi is an almost incredible example of the spectacular success of cooperation between living things. For most of the year, fungi live within the soil as massive networks of filaments, erupting their mushrooms into the open air only to produce their millions of single-celled reproductive spores. Because they have a different physiology from plants (their cell walls are not made of cellulose, like a plant's, but of chitin, the same material as the external skeleton of an insect), they can absorb insoluble minerals and chemicals. These nutrients are essential to plants, and big trees especially have huge demands for them. Plants and fungi have evolved a special partnership, called a *mycorrhiza,* in which the huge tangle of threads of the fungi's body, spreading over very large areas, interconnect with the network of minute root hairs on the feeder roots of plants. It is very much like tapping into one of Nature's Internets, in which a tree can plug into an absorptive surface up to 100,000 times greater than its root system and thereby have access to antibiotics, antiviral compounds, minerals and nutrients for growth, and a much greater water supply. In return, the subterranean fungi receive the by-products of photosynthesis from the plant—sugars and carbohydrates.

These exhibits lead naturally to others dealing with ecological relationships, from life in a British pond to a tropical rain forest. This latter example is expanded in a conservatory that replicates a tropical climate and is explored via curving ramps that ascend through layers of botanical complexity. The conservatory is densely planted with species that illustrate relationships between plants and animals, such as pollination, seed dispersal, and the provision of human foods and medicines. Free-ranging animals here include butterflies and birds, while live fish and reptiles can be seen in underwater views of a flooded South American forest habitat.

Another important exhibition space explores people's relationships with the natural world: our reliance upon it for comfort and survival and the dangers we face by our rapacious interference with it. Visitors are reminded of the spectacle of Nature by a Magic Window observatory, with 3D, large-format HDTV screens presenting apparent real-time experiences unfolding in locations as diverse as an African savanna watering hole or a sub-Antarctic penguin colony. As a conclusion to the tour, a 350-seat theater with an IMAX projection system shows wildlife and environmental films throughout the day.

Wildscreen presents the natural world in fresh ways, and the concept attracted enthusiastic support from local universities, the World Wildlife Fund, the International Union for Conservation of Nature, and the World Conservation Monitoring Center. It should find equal appeal among the public, for it has an extremely rich educational content presented in an essentially entertaining fashion. Its greatest strength lies in its focus on the complexity and dynamics of Nature rather than on isolated animals or plants.

According to E. O. Wilson (1984), a love of Nature is programmed within our genes, through natural selection. He has invented a word for this, "biophilia," which he defines as "the innate tendency to focus on life." Humans seek out Nature because we evolved in Nature. Recent studies have shown interesting glimpses into the ways in which cognitive functions, the ability to focus one's thoughts, and a tendency to seek out new experiences can all be enhanced by interactions with the natural world. Research therapist Bernadine Cimprich looked at recovering breast cancer patients engaged in horticulture therapy programs (Janick, 1992). Those who undertook activities involving direct contact with living plants three times a week were far less likely to suffer mental fatigue, depression, marital problems, or a general inability to cope than those who did not. They also scored higher on cognitive acuity tests and were far more likely to return to full-time work and to seek out new projects such as learning a new language. It is as if, says Yale University professor of environmental psychology Stephen Kaplan (1989) and colleague Rachel Kaplan, "the brain [is] both aroused and soothed by Nature." When Roger Ulrich (1991), associate dean for research at Texas A&M College of Architecture, placed volunteers under stress he found that a view of Nature helped their recovery. He showed his subjects a worker-safety film depicting serious injuries with simulated blood and mutilation. High levels of recovery from physiological stress were achieved with one group af-

Natural history museums have a reputation as dusty enclaves of didacticism, while botanical gardens and zoological parks are typically seen as places principally for recreation. Wildscreen breaks the mold, using various technologies to engage visitors in bright and lively presentations about the interdependencies of all life. Some sections focus on interactive media, others immerse people in replicated rain forest habitats displaying living plants and animals. The moods shift from focused and intense to purely sensory, from intellectual to perceptual. There are good lessons for all natural history institutions here. (© Image Bristol/Virtual Artworks.)

ter they were then shown images of lush vegetation and gentle streams. Blood pressure, pulse rate, and muscle tension all subsided. Individuals in another group that were subsequently shown a film of urban environments recovered either more slowly or not at all during the testing period.

It is only natural, remarks Parsons, for humans to take a close interest in other forms of life. This has been the essence of our survival. The knowledge of which animals could be hunted or should be avoided was once critical to our forebears. Survival depended, too, upon intimate knowledge of which plants and fungi were useful, edible, or dangerous. "Although it is now more important for urban man to be aware of the risks of picking up a power cable than a venomous snake," Parsons suspects that the comparatively small number of generations involved in the transition from hunter-gatherer to worker-commuter means that there is a lot in our genes that still disposes us to be closer to Nature than we realize. All we need, he says, is a little stimulation (Parsons, 1996).

With good fortune, Wildscreen will spawn other such ventures and perhaps be incorporated with existing zoos. It remains to be seen whether people will accept a wildlife center devoid of large animals, but considering Parsons' proven expertise to attract and influence huge audiences with prime-time natural history television documentaries, its future seems secure. In comparison to the dreary standards of most of Britain's zoos, it should surely prove a great success.

This new venture may not display the usual zoo animals, but there are very good reasons why zoos should consider adopting aspects of its philosophies as well as its exhibit strategies and technologies.

GOING DUTCH

Three zoos that have produced ideas or techniques that others should note well are in Holland: at Apeldoorn, Amsterdam, and Emmen. Apenheul Primate Park, in Apeldoorn, specializes exclusively in primates and maintains and displays them in large social groups. Most of them are in display areas that are lacking in design sensitivity, as is typical for European zoos, but at least the spaces for the animals are huge, open, and rugged. A startlingly large number of the animals are also free ranging. Visitors to this unique zoological park get close and personal contacts with the animals unlike anything to be experienced at any other zoo. Perhaps the most refreshing and enjoyable aspects of a visit to Apeldoorn are the sheer sense of *pleasure* that seems to invade the place and the impression that there is enormous happiness among its residents. It demonstrates not only the astounding diversity of primate forms, but something more fundamental, almost edenic: a feeling that we could have had a peaceful relationship with all our other primate cousins had we not chosen to eradicate or subjugate them. There are powerful lessons of wonder and respect to be learned from Apeldoorn.

The zoos at Amsterdam and at Emmen are not specialized as Apeldoorn is. Artis, in the center of Amsterdam, is the oldest in the Netherlands, founded in 1838, and though it is first and foremost a *zoo,* maintaining nine hundred species of animals, it is also a museum of life. Artis has always had strong links with science and includes on its grounds a zoological museum, an aquarium, a geology museum, and a planetarium. In addition it serves as a botanical garden and an outdoor sculpture garden.

Visitors can select from two entrances to this zoo: one is via a planetarium, the other by way of the geological museum. The planetarium emphasizes the fact that whereas our biological beginnings are on Earth, our physical roots are very remote indeed. Every atom in our bodies originated in the interior of a star. We are indeed, as the song says, stardust. Inside stars, the lightest elements, helium and hydrogen, fuse to make heavier elements of oxygen, carbon, nitrogen, sulfur, and phosphorus—the building blocks of life. Those who enter via the geological museum discover equally profound facts about life on Earth. In 1785, when Scottish geologist James Hutton asserted that the Earth was one "super-organism," operating like a gigantic whole creature, he was ridiculed and scorned. Two hundred years later, when American biologist James Lovelock espoused his Gaia hypothesis, a few more people were willing to listen. Slowly, there is growing acceptance of the notion that the organic and inorganic aspects of Nature are inextricably interconnected. A medley of processes maintains an interchange between the atmosphere, soils, and waters and all life on Earth. This story is the common thread throughout the geology museum. It is a lovely introduction to Artis zoo's exhibits of the world's animals. None of these, admittedly, is a good example of zoo exhibit design, but the context in which they are introduced and the museological aspects that are integrated within this zoo provide useful and pertinent lessons for mainstream zoos.

Noorder Dierenpark, in Emmen, is also full of examples of what zoos should be doing. British museologist Kenneth Hudson (1991) has written: "To a person . . . who said he had time to visit only one museum in Europe,

A young spider monkey is introduced to a first-time pregnant gorilla at the Apenheul Primate Park in Apeldoorn: a thoughtful and creative idea to allow her to become used to handling and interacting with baby-size primates. She easily accepted her infant, and then became a perfect example for other females in her group. Since this 1979 event, more than thirty gorilla babies have been born at Apeldoorn, all raised by their mothers. (Courtesy of Apenheul Primate Park.)

I would say, 'Go to Emmen. You will find more new ideas there than in any other European museum. Don't waste your time in Paris.'" It is noteworthy that Hudson refers to this zoo as a museum, for the boundaries between the two are indistinct here.

Opened in 1935 as a private development by the present director's father, Willem Oosting, Noorder Dierenpark enjoyed quite considerable success until the late 1960s, when changing attitudes about zoos across Europe saw attendance beginning to drop dramatically. Oosting's daughter Aleid and her architect husband, Jaap Rensen, shared many of the concerns about zoo standards. They were their own harshest critics. "To put it even stronger," says Aleid, "I hated zoos" (Rensen-Oosting, 1991). The Rensens were faced with few options: sink into bankruptcy, sell the place, or set a new course. They temporarily closed the zoo in 1970 and with loans from the local municipal council set about making essential changes and taking the zoo in a new direction, where education would be its prime purpose. During this period, the animal enclosures were greatly enlarged for larger numbers of fewer species of animals, the infrastructure was improved, and what may have been the world's first zoo education center was built. Noorder Dierenpark undertook a metamorphosis and set itself a pattern of development and level of care that has only kept improving.

Noorder Dierenpark, now run by a private foundation due to the crippling tax restrictions upon family businesses in Holland but still under the management and direction of the Rensens, places strong emphasis on the animals' well-being, with particular consideration for their natural behaviors. The zoo exhibits are not particularly impressive in terms of aesthetics or of naturalism; in fact many of the design details are rather crude, but the animal spaces are functionally sound. An equal concern for visitor comfort is also evident, and creative interpretive exhibits show something rare in zoos—respect for the visitor's willingness to learn.

In many zoos the first exhibit is a flamingo pond: a simplistically bright splash of color and exotica to welcome visitors onto the grounds. At Emmen's zoo the visitor's first encounter is with basalt pillars. The Earth's crust is made up predominantly of basalt and related rocks, and this is where Noorder Dierenpark starts its introduction of the history of life on Earth: inside the zoo's Biochron. Set at the entrance to the zoo, the Biochron is an elegantly designed natural history museum. Formed from the Greek words *bios* and *chronos,* meaning "life" and "time," the Biochron begins where every zoo should, with the story of the evolution of life and an explanation of deep time. Hence the basalt pillars and the adjacent gigantic pieces of granite, sandstone, marble, and other rocks that make up the major part of the continental crusts. These represent that period of time before life forms first appeared, some 3.5 billion years ago, when only rocks and water formed the Earth's surface. A replica of Stanley Miller and Harold Urey's experiment is set up here, wherein the prevailing conditions of the virgin Earth—rain, lightning, and uninhibited solar radiation in an atmosphere rich in ammonia, methane, and hydrogen—were simulated in a laboratory. It was in these conditions that the compounds essential for the formation of living matter reacted, transcending the boundary between inorganic and organic, inanimate and animate.

Instead of flamingos, the first life forms presented to visitors at Emmen's zoo are 2-billion-year-old fossils of cyanobacteria: the blue-green algae that are the oldest producers of oxygen and thus responsible for creating the atmospheric conditions in which the great diversity of life on Earth formed. About 700 million years ago, multicelled creatures evolved from the single-celled bacteria. Sponges, jellyfish, worms, and other simple creatures are exhibited here. Live horseshoe crabs, squids, lobsters, starfish, sea urchins, and other species that have hardly changed for several hundred million years can also be seen, as well as other ancient fishes such as lemon sharks; an interpretation of the history of the development of modern fishes is also provided. A display of tropical fishes known as cichlids, many hundreds of species of which inhabit the Earth's fresh waters, demonstrates fascinating examples of adaptive evolution. When huge lakes formed along the fault line of Africa's Great Rift Valley, about 2 million years ago, some cichlids, probably via floodings and stream runoffs, found their way into these new habitats. There is a wide variety of foodstuffs in these lakes, but only in small quantities. The cichlids adapted and, by natural selection, evolved into many different types, each with the ability to take advantage of a specific type of food and thus occupy a unique niche. (About three hundred species of cichlid have evolved in Lake Tanganyika, and an estimated one thousand in Lake Malawi). They are aquatic parallels to the diversified finches that Darwin found in the Galapagos Islands. Due to their specialized feeding habits—there are snail-eating cichlids, scale-eaters, algae-eaters, and many others—each new cichlid species has characteristically different mouths, like the finches' beaks, perfectly adapted to their individual food specialties.

Before life had evolved onto dry land, some seaweeds in tidal zones had slowly adapted to survive in microhabitats of variably wet and dry environments. It was the beginning of the change from life in the water to life in the dry air. During the Carboniferous period, arthropods such as scorpions, spiders, and centipedes, and insects such as the dragonfly, eventually became the most prevalent animals in luxuriant swamp forests, which are convincingly re-created at Noorder Dierenpark, with examples of the living plant species scarcely unchanged since those far-distant times. Opportunities to tell fascinating side stories are never missed here. A display of spiders illustrates the extraordinary evolution of the spider web. A nearby exhibit of insects notes that even during the Carboniferous era the most common insect was the cockroach.

The end of the Carboniferous age was marked by a shift in the Earth's climate, when it became arid. Swamps dried up, causing many plants and animals to become extinct; reptiles, however, flourished in the new environment and multiplied in form and numbers. Turtles, crocodiles, and tuatara, once contemporaries of the dinosaurs, still exist today. The Biochron displays living crocodiles, agamas, iguanas, turtles, and models of *Tyrannosaurus* and *Pteranodon.* The evolution of turtle shells is nicely interpreted. Although they have existed on Earth for 200 million years, turtles have undergone remarkable developments, altering from a terrestrial to an aquatic lifestyle. The adjustments of the turtle's shell in response to these changes over time are explained and illustrated as an object lesson in adaptation and survival.

Following the abrupt extinction of the dinosaurs, about 65 million years ago, mammals had the chance to develop and diversify. An exhibit in the Biochron explains how they had initially been able to succeed as nocturnal creatures, due to being warm-blooded: a diorama of a small squirrel-like mammals foraging at night next to a sleeping dinosaur depicts a scene from the Mesozoic period, 120 million years ago. Sloths and armadillos, members of the oldest groups of mammals, still retain a nocturnal lifestyle, as do bats, also on display. An exhibit of South American night monkeys explains how these resemble the first primate inhabitants of the Americas. Ancient mammals survived the dinosaur era not only because they could be active at night, but also because of their rapid reproduction abilities, another technique facilitated by the ability to maintain a constant body temperature.

The spectacular and rapid development of mammals in the past 60 million years is the last chapter in the engrossing story told at the Biochron. And at the end, the Biochron considers the finality of our own species; humans are not guaranteed eternal existence. Evolution and extinction have been a normal cycle since life evolved on the planet. Mammal species, for example, exist on average for a few million years before becoming extinct or

The theory of evolution cannot be grasped without an understanding of deep time. This perspective also provides a better vantage for understanding the consequences of our present extinction crisis. The Biochron at Noorder Dierenpark gives visitors the opportunity to consider these fundamental issues, which are invariably ignored in other zoos. (Courtesy of Zoo Emmen.)

evolving into something different. Humans inevitably face one of these alternatives. The last exhibit therefore examines life after humans. It offers the considerations that some remaining animals might adapt to feed upon the abundant detritus of humans' time on Earth, and such possibilities are shown as models: a species of beaver that can consume asphalt as a source of nutrition; worms that are immune to acid rain; and creatures that are able to use isotopes as a food source.

A short slide show before the exit gives a synopsis of the forty-six hundred million years of history that have been covered in the Biochron's exhibits.

It reminds visitors that all of the animals they are about to see at the zoo have not always existed, and will not exist forever; reason enough says the Biochron to examine them carefully.

Considering the local climate, it is appropriate that many of this zoo's displays are inside large buildings. These include a tropical Africa house; a butterfly and hummingbird garden (the only place other than the Arizona-Sonora Desert Museum where hummingbirds breed regularly in captivity); AmeriCasa, where, according to the zoo's guidebook, "a tip of the veil of secrecy that envelops the South American rainforests is lifted" and which includes a nocturnal-animal section and a tropical swamp forest that contains hundreds of piranhas; and a complex of gardens and greenhouses called Hof van Heden (Today's Paradise).

One of the more phenomenal aspects of Noorder Dierenpark is that every two or three years it embarks upon a new interpretive theme. In recent times these have ranged from locomotion to reproduction to colors to odors. Information, though always pertinent, is thus never static, and those fortunate enough to be able to make periodic visits to this zoo can learn more and more about the world of wildlife.

For those who can make only one visit in their lifetime, there are more than enough delights and surprises in this small zoo. A natural-sized herd of hippopotamuses, the largest in any European zoo, reminds one that these are very social animals and that the typical zoo habit of maintaining only a pair of hippos is an inadequate practice. The same philosophy applies to all social species at Noorder Dierenpark. Animals such as ringtail lemurs, colobus monkeys, and hamadryas baboons are maintained here, as in Apeldoorn, in group sizes that not only are natural for the animals but also underscore the drama and spectacle of such species. More than 140 hamadryas baboons, for example, live in the zoo's colony.

The Rensens wanted to create a park where people can feel "respect and friendship" toward other living things and to make it clear to people that they are as much a part of Nature as ants and whales. They aim to provide objective information on the world which their visitors inhabit, and if their exhibits often include humor, that is because "serious subjects particularly deserve that." Just as the natural world would be much healthier if more

There is much to learn about ourselves from observing our biological cousins. This group of hamadryas baboons is part of a 140-member colony at Noorder Dierenpark. Such a concentration might convey impressions to some humans—from the medieval world to the modern, from Brueghel to Buñuel—of a vision of hell. The baboons, conversely, living with so many of their own kind in such good health and comfort, might consider themselves in some sort of paradise. (Courtesy of Zoo Emmen.)

zoos duplicated the Bronx Zoo's efforts in wildlife conservation, so too the zoo world would be immensely improved if more zoos learned from Amsterdam, and especially from Emmen, that a zoo need not restrict itself to zoology, and that the best direction for zoos in the future is to move away from their traditional roles, to embrace new disciplines, new techniques, and new perspectives, and to recognize that learning about Nature can be a most satisfying and useful activity.

A LIVING MUSEUM

Halfway around the world is another atypical zoo, which is nonetheless rated among the top ten zoos in North America and which is consistently reported by zoo directors and curators as their favorite zoo (Gilbert, 1997). The Arizona-Sonora Desert Museum, in Tucson, was founded in 1952 by Bill Carr and Arthur Pack, as a place to teach people the wonders of the Sonora Desert. There are also many guideposts here for zoos and their future.

To arrive at the museum, visitors travel through several miles of desert, immersed, even before they begin their tour, in a world of rocky mountain peaks and gorges and a marvelously rich variety of cholla, prickly pear, ocotillo, palo verde, and other unusual plant forms, most notably the giant saguaro cactus. There is no requirement for visitors to suspend their belief and try to imagine that they are in natural habitat. The crisp light, the susurrant sounds, the characteristic scents of sage and creosote and the constant chattering of cactus wrens, the wide and dramatic vistas—all combine to make a powerful sense of a special place. It is no accident that the icon at this museum is the first vast view that greets visitors at the entrance patio, spreading across a wide valley to mountains in distant Mexico. This was the spot that Bill Carr carefully selected for his new museum.

Carr's move to Tucson from New York for health reasons was an auspicious relocation. He had earlier developed his regional exhibit ideas at the Bear Mountain State Park, a trailside museum along the Hudson River, but immediately fell in love with the desert and was disappointed to discover that his enthusiasm was not often shared by fellow Tucsonans. He reasoned that their attitudes and destructive habits might change if they could have meaningful and intimate experiences with the land and its plants and animals. With financial backing and enthusiastic support from conservationist Arthur Pack, publisher of *Nature* magazine, his idea of a museum as a living, outdoor educational center—a microcosm of the Sonora Desert—became reality.

During its early years the exhibits were simple, but in the late 1960s and early 1970s Merv Larson, first as exhibits curator and later as director, created exhibition standards that still stand today as some of the finest replicas of natural habitats to be found. An unusual mix of artist, builder, and zoologist, Larson secured the museum's place on the international zoo/museum map, with exhibits of unusually realistic appearance for small cats, otters, beavers, and bighorn sheep. Earlier, Larson, Carr, and museum director Bill Woodin had created an equally wondrous life-underground exhibit. Built on a tight budget, it never included all the space and ideas of its designers, but nonetheless revealed a hidden world that found great admiration among both the public and the profession and established the museum's tradition for innovation and creativity. Larson's most impressive triumph was his replication of a full-size limestone cave, so realistic, so amazingly perfect in detail, it still convinces visitors that it is a natural geological feature of the site. His creative genius, however, did not depend upon such temporal confinements as projected work schedules, closely watched deadlines, and bottom-line budgeting. He was producing a work of art. The board of trustees, restricted to a business perspective, could not understand or tolerate this untidy approach. Larson was taken on a walk around the grounds one morning by board president Bazy Tankersley. By the end of the walk he was no longer director.

From then, and through the 1980s, the museum began a persistent decline, building a mountain habitat of merely average zoo standards, a consciously architectonic hummingbird aviary located on too prominent a knoll, and a walk-through aviary which, though delightfully landscaped by the museum staff, is a cumbersomely engineered structure similarly clumsily sited. Years earlier, Larson had sketched plans for an aviary to be located within the depression of a desert wash, so that tall side-supporting structural members would not be required. It was the exact opposite of what was built after his rapid departure. His design standards were not pursued, nor his creative ideas, although he was working just a few miles away, establishing the Larson Company, which is now responsible for zoo design and exhibit construction around the world.

Larson's successor, Holt Bodinson, with a degree in business administration, stayed just one year: long enough to preside over the resignation of nine senior staff and all the volunteer docents. Suddenly it seemed that the museum was about to sink. Dan Davis, the new director in 1978, managed the Herculean task of righting this badly listing ship during his ten-year

"Life Underground" was one of the earliest examples of the innovative displays that have characterized the growth of the Arizona-Sonora Desert Museum in its fifty years of often progressive development. (Author photo.)

tenure, but his expertise was not in exhibitry and interpretation. His instructions for the new mountain habitat, for example, were to make it of generic rock formations. The lessons of exquisite attention to detail of the museum's cave were learned only in regard to their negative aspects of cost and time.

There are interesting revelations to be found in this mixed history. From the beginning, visitors have extolled the naturalism of the Desert Museum and talked with enthusiasm about its natural habitat interpretations. Yet for the greater part of its history, the museum has not provided very satisfactory housing from the animals' point of view. The first exhibits were miserably small cages in long, straight rows, the animals in tidy taxonomic packages. Even today not all the basic problems have been resolved. The small cats can never get distant views from their enclosures, though this is an essential and basic need for most such predators. These enclosures are nonetheless continually praised because they look realistic to visitors. There still exist, too, some woefully inadequate, small, round cages for birds of prey, too small even for the many that have been injured in the wild and are unable to fly; until recently a jaguar lived in a naked cage unworthy of even the shabbiest small city zoo; and a pair of black bears exist within a sterile grotto in the mountain habitat. These inadequacies, however, do not seem to upset people, and it is illuminating to examine the reasons. The great majority of visitors to the Desert Museum are on vacation, elated about being in pleasant warm sunshine in the middle of winter, immersed in the natural beauty of the desert with wide open vistas of stunning beauty around them. They feel good. No wonder they like the museum and accept it so unquestioningly. The same phenomenon occurs at the San Diego Zoo. The abundance of inadequate exhibits at San Diego would not be tolerated in a different environment. Zoos in places such as Detroit, Boston, or Philadelphia would be (and have been) publicly chastised for similar exhibit standards. But visitors to those northern zoos are not under San Diego's blue skies, surrounded by hibiscus and palm trees. Most people judge zoos not from the animals' criteria but from selfish reasons.

A TALE OF TWO OTTERS

About twenty years ago, the river otters at the Desert Museum so inspired Lewis Thomas that he wrote of the experience in his book *The Medusa and the Snail.* "I was transfixed. Swept off my feet; I wanted never to know how they performed their marvels; I wished for no news about the physiology of their breathing, the coordination of their muscles, their vision, their endocrine systems, their digestive tracts." Thomas says that what exulted him was not that he had learned anything new about otters, but about himself, and "maybe about human beings at large." He learned about "surprised affection," and drew lessons from the experience about the strength of our needs for friendship.

There are, however, other equally illuminating lessons to extract. Delightful and heartwarming as Thomas's account is, it is salutary to reflect that this uplifting experience was achieved at the price of having otters in captivity. This in itself is not necessarily an abuse. Otters in captivity can enjoy a life of greater comfort and duration than in the harsh reality of the wild.

They can enjoy freedoms unknown to wild animals: freedom from hunger and from parasitic debilitation, freedom from hunters and other predators, and from untreated illnesses. But the otters at the Desert Museum, until very recently, spent their lives trapped in what was essentially a pit. The naturalism of the enclosing walls was deceptive to human eyes, but in the animals' daily routines they never escaped above it, never saw beyond its enclosing rim.

Two specific points worthy of reflection emerge from this experience. One is obvious, and one we must choose to take. First is the clear moral that zoos are places built for human indulgence. We keep animals in captivity not, as some zoo apologists insist, for the well-being and safety of the animal, but for our own satisfaction, curiosity, enjoyment, and, in good zoos, for lessons about life. That we can do this and at the same time (when we do it right) provide wild animals with safe and contented lives is fortunate happenstance. The second point to be considered is that we should learn to measure the quality of zoo exhibits not from a human perspective but from the degree to which the animals' needs are met.

No visitor ever complained to the museum's management that the otter in the enclosure at the Desert Museum did not have a distant view out of and beyond the high edge of the enclosure, but the complaint that there is only one otter is persistent. "He must," visitors say, "be so lonely." In reality, however, North American river otters are not social creatures. Apart from brief times of mating and when a female is raising young, river otters (like many predators) are solitary hunters. On very rare occasions siblings will remain together as a pair, but this is an exception. Humans, being social creatures, think a solitary life is unnatural, and so it is for many species, but by no means all. Some visitors say they would prefer to see a whole group of otters: "They are so playful, surely they would be happier?" From a river otter's perspective, few situations could be more stressful. If we are to base our assessments of zoo exhibits on animal needs, it must, then, be done not from misguided anthropomorphism, but in light of the differing and specific natural requirements of each species. Oftentimes, this would require visitors to look for provisions in the zoo enclosure that might be contrary to their human, personal wishes. Do the animals have the ability to get out of view of people if they so choose? Do they have space to get away from close proximity to visitors? Can they hide from other animals in their enclosures if they so choose? Do they have access to shady places in hot weather, warm spots on colder days? Are they able to enjoy distant views outside their enclosures? Is there a range of natural objects, materials, and vegetation to interact with? In short, do they have opportunities to carry out the important aspects of their natural lifestyles?

Some favorite zoo animals are probably not suited for life in captivity; bears, for example, are creatures of such strength and dexterity, and have such a capacity for rooting about, exploring, and destroying, that meeting their behavioral needs in a zoo enclosure is extraordinarily difficult and time consuming. Elephants are not good candidates for zoo life because their needs for space, environmental complexity, and social interaction are beyond anything that most zoos can provide. It is rare to find a zoo elephant that is in very good physical or psychological health. Killer whales and dolphins cannot enjoy a full life in their inevitably small and shallow pools. Reptiles are frequently maintained in exhibits so small that there is not sufficient room for snakes to stretch their bodies to full length or to have access to varying microhabitats. In many zoos they rarely have opportunities to explore a different spot from the depression they lie within all day and night.

Conversely, some species that elicit human sympathy can be fairly easily accommodated in captivity. Primates are the most obvious example. With good will and ingenuity it is not especially difficult to provide an exhibition habitat that can keep them occupied and active. It is, as Heini Hediger pointed out, a matter of the quality rather than the mere quantity of space for such species. A curious incident at London Zoo in the winter of 1912 sheds some light on the matter. Jacob, an adult male orangutan, smashed a skylight and escaped from the ape house. Interestingly, he did not then set off across the park but simply climbed into a nearby lime tree, broke off some branches to make a nest, and settled down to sleep for the night. Next morning he was coaxed back into his cage. Jacob was not seeking freedom, but merely the chance to make a bed of natural foliage, rather than having to sleep on a hard, flat floor.

To create satisfactory captive conditions it is necessary to satisfy the fullest possible complement of each species' social, behavioral, physiological, and psychological requirements. In recent years the Desert Museum has been recognizing such needs, not just for mammals and birds, but also the snakes and amphibians and even the invertebrates in its collections. The otter, too, now has an enlarged enclosure, providing not only views to the outside desert and mountains but also a natural substrate, live vegetation to interact with, muddy areas, and pools.

EMBRACING THE DESERT

It is curious that in the first forty years or more of its life, the Desert Museum never actually developed any essentially desert landscape exhibits. It is perhaps the reason that the museum has not had the eventual effect that Bill Carr had hoped, of encouraging Tucsonans to stop their habit of desecrating the natural desert. Tucson, having despoiled its original nest, is now spreading like a voracious cancer across massive tracts of what was beautiful landscape and rich wildlife habitat. Early in its history, the Desert Museum fell into the trap of so many other zoos and allowed itself to be lured more by exotica than by the local. Thus, its riparian beaver and otter exhibits, the bear and mountain lion of the mountain habitat, the prairie dogs, the jaguar—none of these are fundamentally representative of desert.

Although not all other zoos can present such awesome vistas around their sites or have their visitors so thoroughly immersed in the landscape they are interpreting, there remain critical points at the Desert Museum that all zoos should consider for the benefit of their futures. Each of these is illustrated within a new Desert Trail exhibit, designed by staff landscape architect (and herpetologist) Kenneth Stockton.

First, of course, is the fact that every natural feature within this exhibit is completely authentic. Rocks, trees, and shrubs have been added, but they are all taken from and completely natural to the desert habitat that the museum interprets. The naturalness, however, is enhanced by placing many of the more interesting plants and other natural features close to visitor pathways. It is also kept pure by the invisibility of the enclosures for the animals. Stockton has developed a patented wire mesh that is so gauzy yet so strong that it can safely contain strong and destructive animals while being virtually undetectable, even within the diaphanous desert landscape. Placed against even sparse vegetation, it cannot be detected. The fence posts that hold this "Invisinet" would be prominent and visible features, but Stockton

The value of trying to create the illusion of wilderness in all details is neatly demonstrated in this example from the Arizona-Sonora Desert Museum. In a recently constructed javelina exhibition habitat, designed by landscape architect Kenneth Stockton, there is apparently no barrier between people and animals. The sense of discovery and excitement is thus greatly heightened and the experience has more meaning. (Photo courtesy of Kenneth Stockton.)

has devised various cunning alternatives. Metal structures that resemble dead branches or cholla cactus skeletons hold the mesh in place. Taller metal support posts are clad in saguaro ribs, so they exactly resemble the saguaro skeletons that occur in the natural landscape. The effect is that visitors discover animals in their natural habitat, as if by accident. It creates wonder, amazement, and sometimes concern for the visitors' own safety. With no visible demarcation between visitors and animals, the sensation of being inside a wild animal's habitat is intensified by the fact that visitors are not simply looking into an enclosure from selected viewpoints. In one section of the trail, a herd of javalina are all around the visitors, because the javalinas' territory goes under the public route in several places, crossed by what appear to be rustic and old stone bridges.

Obvious changes in both visitor and animal behavior have occurred since the javalinas were relocated to this new area, similar to those that occurred with the new gorilla exhibition habitat at Woodland Park Zoo. The animals spend much more time in vocal communication, muttering contentedly as they appear and disappear from each other's sight. They are more active, in a pleasant and gentle way, more social and playful, and, thanks to the addition of seeps in the area, often muddy from snout to tail. The visitors, meanwhile, are much more attentive to all this activity. They appear more alert and more surprised and spend far longer exploring and watching than they did at the walled enclosure of the old and very much smaller exhibit. The skill with which Stockton has created a stage for natural drama is impressive and has achieved precisely the goal that all zoo exhibits should have. His painstaking and meticulous concern for detail has resulted in an exhibition habitat that precisely matches all aspects of the wild, with no hint of artificial components to destroy the visual illusion.

The second important aspect is the size and diversity of the spaces provided for the animals. It may not prove to be especially easy for visitors to see all the coyotes, mule deer, javalina, jackrabbits, lizards, roadrunners, and other animals in the enclosures of the new Desert Trail exhibit, at least not so conveniently as in their old cages and small pens. But when the animals are spotted, the experience for visitors is as exciting and as authentic as the adrenaline rush one gets from suddenly catching sight of animals in the wild. Convenience and obviousness have been swapped for experiential authenticity. For the animals, the variation in types and levels of terrain within these large enclosures bring the benefits of diversity and change. Here, ease of keeper maintenance has been traded for animal well-being.

The third lesson for zoos in the Desert Trail exhibit is the fact that it is about far more than the animals. The display and interpretation of the characteristic plants of this region are given equal attention, and especially the interrelationships and interdependencies between plants and animals. Convergent evolution is interpreted with examples of flora and fauna that have independently evolved very similar characteristics. Sphinx moths and hummingbirds, for example, have each developed long tongues and hovering flight and thus pollinate deep-throated flowers from which only they can feed. The exhibit area is also scattered with examples that interpret the ways in which plants have evolved for survival in the extreme dryness of the desert. One small exhibit reveals how some tropical forest plants have adapted to become desert dwellers and are now very different from their original forms. Further, there is close attention to the diverse small animal populations of this desert. An enclosure for lizards is as large and more varied than some zoos give to lions. A special section displays animals as tiny as scorpions and spiders and other diminutive animals that shelter within the fissures and crevices of rocks in the desert. These, too, are shown in their natural micro-habitats, within an eroded cliff face of ultra-realistic rockwork. The entire Desert Trail exhibit is, says Stockton, "like a giant canvas that allows constant additions to the story of the desert."

In the almost perpetually blue sky over the museum, visitors can see soaring raptors almost any day. Their appearance, however, is sporadic: they are just riding the thermals, scanning the desert floor like a stockbroker looking for pickings in the business section of the newspaper. To guarantee that their visitors can see birds doing something interesting, many zoos have built amphitheaters, where they present trained-bird shows. These are extremely popular with zoo visitors, particularly because they ensure some action. Unfortunately, there is usually more activity from the presenters than the birds, and the common intrusion of amplified music is hardly appropriate for encouraging thoughtful consideration of bird behaviors. Even if these excesses could be deleted, however, there remains the problem that the birds in such a staged show are inevitably presented out of context. At the Desert Museum, a system has been devised that allows visitors to witness the drama of birds of prey in natural behaviors and to see these spectacles in the birds' natural habitat.

A family trio of Harris's hawks has been conditioned to fly free across the Desert Trails exhibit and demonstrate their cooperative hunting techniques, but to then return to their trainer. The birds swoop and rush through desert washes, occasionally in real pursuit of a wild rabbit but usually doing it only for the reward they know will be forthcoming from their handler. Their arrival in the habitat is not announced or promoted. Released from a distant site, they simply appear, as they would in the wilderness if one were fortunate enough to be there. Volunteer docents, who are always out and about the grounds, interpret the behaviors spontaneously to visitors in the vicinity. At present the Harris's hawks are the only species trained for these activities, since the flights were a limited experiment in light of concerns by board trustees who questioned its cost and effectiveness. Fortunately, public response has been so positive that more such flight programs may be attempted in the future. One very inexpensive idea is to attract local wild birds to assist in such a demonstration, by collecting road-killed animals, currently gathered by local park officials for incineration, and placing them out in the exhibit area to attract wild turkey vultures. It would provide a good opportunity to interpret how these birds perform such useful tasks as recycling carcasses.

In addition to the Desert Trail's interpretations of zoology and botany are displays dealing with paleontology and archaeology. As with Noorder Dierenpark, the intent is to give visitors a sense of the deepness of past time (tens of millions of years since *Sonorasaurus,* whose skeleton is on display here, fell into its muddy grave) and of the long history of interaction that people have had with the desert environment. A replica of prehistoric Hohokam agave fields reminds visitors that people have lived successfully in this desert for many thousands of years, without supermarkets or air-conditioning and without destroying the desert's resources as have its current residents. Tohono O'odham Indians generously give seasonal demonstrations of saguaro fruit harvesting, basket weaving, and other traditional skills, reinforcing messages of learning to embrace the riches to be found within the desert. With occasional frugality, care, and foresight, people have lived in this spot in the desert continuously for more than ten thousand years. With more care and foresight, humans could continue living there for thousands more—but the current systems are not sustainable.

The mix of stories and displays about the mammals, birds, fish, reptiles, invertebrates, peoples, plants, and all their interactions, as well as the geology and deep history of the region, makes for a powerful and seductive vehicle to interpret the desert. It is exactly what zoos need to do if their messages about the wonders of life on Earth, and our wise stewardship of it, are to be effective. The Arizona-Sonora Desert Museum and the Noorder Dierenpark are two private nonprofit institutions that have shown the effectiveness of this approach, one regionally and one worldwide.

THE REGIONAL VIEW

The concept of the Desert Museum as a regionally based facility attracts much admiration within the profession but, perversely, the concept has not spread. As John Grandy, vice president for Wildlife and Habitat Protection for the Humane Society of the United States, has said (1992), there should be "six to twelve regional bioparks where meaningful educational material is available and where native wildlife can be studied and appreciated in their natural habitat." Yet few such institutions have been established: the Oregon High Desert Museum, in Bend; Northwest Trek, in Washington State; the Desert Wildlife Park in Alice Springs and Healesville Sanctuary, both in Australia; and California's superb Monterey Bay Aquarium, marvelous in the original sense of the word. A few zoos have also developed specific exhibits on their own bioregion, and they each reveal the high order of veracity that can be achieved with this approach. Witness the Northern Trails exhibit at Woodland Park Zoo, the polar bear display at Point Defiance Zoo, the South Carolina Gallery of Riverbanks Zoo.

A Louisiana swamp exhibit that opened in 1984 at Audubon Park Zoo, New Orleans (Burnette, 1986), blends both natural and cultural history in a five-acre replication of the swamplands of southern Louisiana. It is a good example of authenticity and shows how convincing the experience can be when the focus is on the local region. Visitors are deeply immersed in the heat, humidity, birdcalls, insect life, smudgy light, scents, and sensations of this dark and mysterious habitat. Following trails of crushed shells, wooden bridges, floating docks, and cypress boardwalks, they discover and learn about the human, animal, and botanical swamp dwellers. And always they are surrounded by dense, thick vegetation. Green living things soar to the sky, festoon from overhead, float on and through the water, crawl over the buildings, writhe along the paths, wrap themselves around your legs if you stand still too long, and demand constant attention from zoo staff to maintain the mood of a real swamp while still allowing comfortable access.

The nearly vanished folkways and folklore of this bioregion are woven through the exhibit. Trappers' cabins, skiffs, beehives, traps, fishing paraphernalia, traditional crafts, and occupations are illustrated and interpreted, like a time capsule. In a lagoon green with duckweed and populated by alligators, an old houseboat is moored to a rickety fishing pier. Some of the other exhibit areas feature now rare animals that once were common here, like cougars, or species that have been introduced and become feral, such as nutria and water hyacinth. There are also black bears, lots of different turtles, raccoons, otters, armadillos, and swamp rabbits, which people have to look to find. A bird lagoon is inhabited by great blue herons, egrets, redhead ducks, white ibis, night herons, cormorants, all seen and thus better understood because they are in the context of their native home. A mock-up of an old trading post, festooned with Spanish moss hung out to dry, provides opportunity for interior displays of frogs, fishes, and reptiles in glass tanks, illustrating the many aquatic microhabitats of the swamp. A café in vernacular style, halfway along the trail, continues the authenticity of the experience, offering jambalaya, gumbo, étouffée, and red beans and rice, which people can enjoy while sitting on a wooden deck, beneath the shade of bald cypresses, overlooking an alligator pool, while listening to Cajun music.

Around the world, other bioregions of equally wonderful complexity sadly remain uninterpreted in any unified way. They are shown in an atlas produced by the International Union for the Conservation of Nature that divides the world into biotic provinces. The natural patterns on this map are starkly in conflict with the political construct of states that politicians and soldiers have created with mathematically precise boundaries and unnatural geometries. It makes one realize that in the natural world there is no such place as Oregon, Texas, or Ohio. Or Chad, or Albania, or any of the other places created by artificial demarcation. If humans understood and respected the watersheds, landforms, plant associations, soil types, and other natural components of a bioregion, then residents of North America might say that they lived in Ozarkia, Shasta, Cascadia, Katuah, or some other naturally identifiable region. Children would study their home place as a complete and natural entity, and laws would be written to protect the integrity of these places, simply because they had been defined and named as such.

Every person knows when he or she is in a particular bioregion. The place sounds and looks and feels different from any other. And people instinctively know when they have passed from one such place into another, in a way that passing a state boundary marker on the highway can never achieve.

The Sonora Desert is celebrated by its own museum. There should be a similar museum for Cascadia, one of the richest and most spectacular biomes of North America, yet not one institution displays and interprets this wonderland. The Rocky Mountains region is equally neglected, as are the short and the tall grass prairies, the taiga and tundra of Alaska, the temperate hardwood forests of the north-central and eastern states. Each of these regions deserves its own interpretive museum. The same is true in many other regions around the world. There is no natural history institution in Britain dedicated to its habitats, geology, wild animals, and botany. Nor is there one in any other region in Europe. The absence of such interpretive institutions mirrors our ignorance of and subsequent irresponsibility to the land (and the sea). Aldo Leopold (1966) theorized that humans abuse the land because they regard it as a commodity that they own: "When we see land as a community to which we belong, we may begin to use it with love and respect." We might come to recognize the need for us to commit to a place.

NATURE IN A BOX OR A BOTTLE

Although the idea of bioregional parks has not yet flourished, a rash of other wild-animal shows has erupted in very unlikely and unpleasant spots. At retail complexes around the world, from Tokyo to Las Vegas, entrepreneurs are luring shoppers inside with mini-zoo experiences. It seems that many people want their taste of Nature in small and homogenized doses, and the Ogden Corporation of New York is one of the business organizations capitalizing on this, building "American Wilderness Experiences" in, God forgive them, America's shopping malls. The first opened in 1997 in the hyperactive sprawl of Ontario Mills, at the junction of Interstate Highways 10 and 15 on the dusty outskirts of Los Angeles. It is set between Game Works, a virtual-reality arcade, and Dave & Buster's, a combination restaurant and game arcade.

A cardboard "forest" welcomes the ten-dollar-paying visitor, where hidden canisters imbue the air with an unnatural scent. So-called rangers, who traverse the approximate fifty yards from one end of the Wilderness Experience to the other, between the Naturally Untamed clothing boutique at the beginning and the faux log cabin restaurant at the end, give presentations that vary from a weary monotone to feverish interaction with visitors. Before entering what Ogden has the gall to call their Biomes, visitors must endure a wild-ride simulator theater, where both the patronizing commentary and the jarring, plunging seats can incite nausea.

The biomes of North America, those natural regions that so deserve to be celebrated, interpreted, *understood,* are reduced by Ogden to five rooms of artificial rockwork, designed by Patrick McBride, a Florida retail architect whose credits include the Hard Rock Café and the gift shop at Biosphere 2. He describes himself as an expert in "shoppertainment." The animals on display are confined to tiny and useless spaces behind glass panels. These cages have their own air circulation system, so that animal scents won't compete with the smell of cooked food in the Wilderness Grill.

F. William Ziegler, the grandly named vice president of Science and Conservation for the Ogden Corporation, believes for some reason that this ventilation system, plus lighting that simulates dawn and twilight, "approximate[s] the animals' outdoor habitats." Pointing out that "Modern consumers yearn to get back to nature but don't have the time," Jonathan Stern, originator of the Experience, says that "Parents are dying for some place educational and fun to take their kids, and to do it in a relatively short period of time" (Bird, 1997). He envisages fifteen more Wilderness Experiences across North America and at least the same number overseas, with maybe one hundred mini-versions in smaller malls across the country. In 1997 Ogden had more than $100 million committed to its first eight sites. It is hard to imagine that even a populace that loves malls could devote time and money to such a miserably false experience. The chances of commercial success seem remote. And since Ogden is in this venture only for the money, not for some compulsive desire to promote wildlife conservation, then the whole idea will probably be dumped.

Public aquariums have proven themselves enormously popular, and in places as different as Chattanooga and Baltimore have proved their ability to enliven and energize dead areas of a city, such is the strength of their appeal to middle-class families. Planet Ocean is a business venture that now plans to open aquariums around the country, starting in Seattle, Minneapolis, Chicago, and New York. Admission to these aquariums will be free, for they serve only as backdrops to gift shops and to restaurants seating 250 to 500 diners. It claims to be "A Company with a Deep Philosophy and Commitment," although it seems that the philosophy is dollar based and the commitment is to profits. Their promotional brochure asks whether there is "a better way to promote environmental responsibility and educate families about the ocean than by designing a restaurant/retail company around it?" The question is presumably meant to be rhetorical, but it begs the obvious response that there are indeed *much* better ways. In fact, reducing the ocean environments to a mere background for dining and shopping is likely to prove a disservice to any conservation ethic. How can one form respect for any wilderness habitat when it is debased to a themed shopping backdrop?

Ocean sound tracks and "salt water breezes" will fill the restaurants at Planet Ocean. Drinks at the bar will have names based on ocean themes. It promotes itself as "an experience unlike anything you ever fathomed." Unfortunately this type of commercial opportunism has long existed. There is a seamless kinship between the shambling, dancing bear on a rope in medieval fairs, through the traveling menageries of eighteenth-century Europe, to the shabby roadside zoos at gas stations along America's first highways, to the glitz, the artificial adventurism, and the exploitation of Planet Ocean.

Some will reject this criticism, claiming that the Planet Ocean and American Wilderness Experience ventures are just clever marketing strategies and maybe even an inevitable evolution of zoos. Ever increasing pressures to make money are now driving the decisions for many zoos. When a major zoo recently purchased nine gorillas it was promoted as a strategy to help save gorillas, but was in truth aimed at increasing attendance. But zoos have always been a business as much as anything else. Some have simply erred more in favor of placing profits as their principal purpose. Anheuser-Busch, the world's biggest brewer, does not operate some of the largest zoos in America because it is committed to conservation or education. The corporation may promote its parks as "magical," places "to make contact . . . to bring our world together" (Davis, 1995), but the business imperatives are surely more fundamental. If these places did not contribute to the corporate bottom line they would be off-loaded. However, in assessing such strictly commercial ventures as American Wilderness Experience, Planet Ocean, or Sea World, one must differentiate between hard-edged business tactics that are crass and those that only seem so because they are bold and new. Much of the current publishing world, for example, layered with marketing specialists, sales executives, and business managers, tends to regard books not as intrinsically worthy products that deserve to be produced as vehicles for ideas, but as cogs in a merchandising machine that link the book to the television miniseries to the magazine feature to the movie to the videocassette to the soundtrack CD to the action toys to the kid's pajamas.

MICKEY GOES TO THE ZOO

What, then, are we to make of the Disney corporation's latest and biggest and essentially commercial adventure? In April 1998 Disney's Animal Kingdom opened, in Orlando, Florida, after eight years of planning, hundreds of thousands of miles of travel by staff and consultants to all parts of the world, placement of more than 4 million plants and sixty miles of pipe and six hundred miles of electric cable in the ground, signatures on more than five thousand employee contracts, transport of 1.5 million cubic yards of soil, and, most staggering of all, the reputed expenditure of far more than one billion dollars.

The promotional materials for Disney's Animal Kingdom repeatedly stress that the purpose behind this vast effort "is to inspire a love of animals and concern for their welfare." It is not mere cynicism, though, that suggests there might be more prosaic purposes overlying this worthy aim. Cross marketing of Disney's animal brands springs to mind as one strong motivation. And in synergy with all the other Disney developments in Orlando—Magic Kingdom, Disney-MGM Studios, Epcot, three water parks, sixteen hotels, six golf courses, Disney's Wide World of Sports complex, the Pleasure Island village of nightclubs, and the Downtown Disney megamart—Disney's Animal Kingdom will surely help extend the current average three-day tourist visit, increasing overall operating profits and the value of Disney stock.

Disney's Animal Kingdom arrived on the scene after a bumpy ride for the corporation through the 1990s. Euro Disney (renamed Disneyland Paris

Resort) had been not so much a coup for Walt's descendants, more a *manquer son coup*. Disney's America, a historical theme park for northern Virginia, was even more ill conceived and died before it drew breath. Disney hopes to reverse its fortunes in the next decade, however, and is betting on the beasts to do it.

Noorder Dierenpark and the Arizona-Sonora Desert Museum are small institutions that have been carefully guided through critical phases by just a few dedicated individuals, evolving slowly over several years of patient and steady growth. Unfortunately they have been ignored as role models by other zoos. Disney's Animal Kingdom, conversely, has burst upon the zoo scene and is so massive and is attracting such huge audiences that other zoos will find it impossible to ignore. Replete with enormous resources, boasting almost incredible statistics, and planned by a veritable army of experts, some wonder if this is the ultimate zoo. Several zoo professionals query whether there can be further evolution of zoos after this gigantic effort. Most certainly there will be, particularly in quality of experience and depth of messaging. Disney's Animal Kingdom is impressive in many ways, but its lack of innovation is disappointing. Indeed, the size and speed of this project have hampered its capacity for originality and thoughtfulness.

Everything about the Disney corporation is big. In Glendale, California, Disney maintains a think tank that employs almost two thousand designers from virtually every design discipline. There are engineers with a vast range of backgrounds and experiences, planners, sculptors, artists, computer wizards, mechanics, and creative whiz kids. They are called Imagineers, and they call their place of work The Dream Factory. They work for a corporation that David Remnick, in a *New Yorker* article on Disney (1997), accurately describes as "desperate to understand and profit from an era that promises unprecedented time and resources for the amusement of sedentary human beings." The Imagineers exploit all the newest technologies and are engaged with projects as diverse as virtual-reality attractions, dancing robots, the California Adventure theme park that is to be built in the Disneyland parking lot at a cost of two billion dollars, fireworks systems that don't emit smoke, digital movie projectors, and a new aquatic theme park called Tokyo DisneySea. A few years ago some Imagineers were producing concept sketches and ideas for the Animal Kingdom project. They developed such things as an artificial baobab tree that includes stainless steel, giant-sized lazy

Susans to feed giraffes on a regular basis throughout the viewing day, and, for the park's central icon, a Tree of Life that is 145 feet high and 170 feet in circumference, with a swirling tapestry of hundreds of carved animals emerging from its trunk on the outside and a 3D movie theater in the interior.

Clearly, other zoos cannot duplicate such wizardry or complexity. Nor can they copy Disney's scale or its staggering statistics. The new park, covering 500 acres (twice the size of Epcot), is planted with 100,000 trees of 850 species, 250 species of vines, 1,800 species of ferns, 260 different types of grasses, and 4 million shrubs. Forty thousand mature trees were transplanted, including one that tipped the scales at ninety tons. The park even has its own eight-acre browse farm of trees and shrubs, to provide animal exhibit areas with a constant flow of replaceable forage (Malmberg, 1998).

But other zoos can learn from and then aim to improve upon Disney's success as a storyteller and with care and time could be far more convincing. Disney's plots have rarely strayed from shallow waters, and at Disney's Animal Kingdom conservation messages become little more than aphorisms. Even so, Bran Ferren, who directs Disney's research and development, and who recognizes that raconteurs are the world's oldest professionals, asks, "Do you ever find yourself saying, 'The problem with television is that it needs four hundred and twenty-two more horizontal scanning lines'? No. The problem with TV has to do with story." And most stories, he says, "really suck" (Remnick, 1997). Many of the exhibits at Disney's Animal Kingdom suffer from the same condition. They have the most perfect cast here, in the form of living wild animals, the best material in the world, the most potentially astounding plots, but like most zoos, they rarely take the risk of engaging their visitors in much beyond the superficial.

Rides varying from serene to suicidal are scattered throughout Disney's Animal Kingdom. Shows include an inevitable "Journey into Jungle Book." At Camp Minnie-Mickey one can see the Festival of the Lion King or watch trained birds at the "Flights of Wonder." "Exploration Trails" wander through misted gardens of prehistoric plants or an "Oasis" of flower gardens, waterfalls, and grottos populated only with pretty and cute animals. DinoLand USA includes a gigantic playground and a Restaurantosaurus. A petting zoo, reached via the Wildlife Express train, is called, unblushingly, the Affection Section. Musicians from a multitude of world cultures meander through the park, strumming, humming, singing, and blowing. An abundance of shops and restaurants skillfully vacuum the pocketbooks of visitors to Harambe, an ersatz African market town. It can all combine to be daunting, and wearying, and the visitors have not yet reached the park's main attraction, Kilimanjaro Safari.

Skills, patience, and budgets were stretched to breaking point in putting this huge project together in such a hurry, and unique construction problems paved the whole journey. The fourteen-story-tall Tree of Life has to sway like a giant tree, yet withstand the ninety-mile-per-hour winds that sometimes scour central Florida. Designed on oil-rig structural technology and bolted together in gigantic pieces, it nonetheless maintains a sense of fragility and intricacy, and may be the most fabulous thing Disney has ever made, at least since Fantasia. Joe Rohde, the park's executive designer, wanted this icon to be a celebration of emotions about animals and their habitats. Carvings of more than three hundred animals, sculpted by a dozen artists from around the world, emerge from the bark in a balance of animal forms and wood textures. Inside is a 450-seat theater for a 3D movie, laden with special effects, about the bad treatment insects receive at human hands. Outside, the tree's giant roots twist over and into the earth, melding with a landscape of pools, meadows, and forest groves. Here are exhibition habitats for flamingos, otters, capybaras (145-pound rodents), lemurs, axis deer, giant tortoises, wallabies, green-winged macaws: a selection of species carefully limited to the weird and the cute.

Similarly, the Oasis, the opening scene of Disney's Animal Kingdom experience, displays only odd or colorful animals in a peculiarly Hollywood version of Eden. Tree kangaroos, muntjacs (miniature deer), giant anteaters, hyacinth macaws, and other brightly plumed birds are found amidst gardens full of colorful flowers, waterfalls, and thick green vegetation, all intended to be a soothing introductory antidote to the advertising hyperbole that has lured visitors to the park. Unfortunately, the animal exhibits at the Oasis and around the Tree of Life are decidedly ordinary, formed with the usual lumpen and illogical masses of incompetent fake rock forms, from no fixed abode, that stumble all over zoos around the world.

At Conservation Station the accent is on statistics and information rather than impressionistic experience. Even the bathroom stalls have Poop Facts on the back of the stall doors. Computers print out lists of conservation organizations and related programs for kids in their hometowns. A mermaid puppet recounts tales of children who have been directly involved in conservation efforts such as beach cleanups, organizing a March for Manatees, waging letter campaigns to save dolphins from tuna nets. The veterinary offices are located here, and are not only open to public view but wired for sound, so that visitors can watch and talk to the veterinarians at work, carrying out anything from regular health checks for a turtle to a root canal on a lion.

The nearby petting zoo houses domestic representatives of rare and endangered breeds. Indeed, scattered throughout this peculiar amalgam of a zoological garden and recreation park are instances that suggest a desire to make people aware of the plight of animals. Rick Barongi, director of animal programs at the park, explains that they "started with a strong philosophical message of conservation, and added to it the thrill of adventure" (Corliss, 1998). The difficulty for Disney will be to manage a sensible balance between the two. The potential for conflict, for the spectacular and sensational to overcome the message, is quite real.

The discordant mix of objectives is exemplified at DinoLand. A fossil preparation laboratory gives a view of a dinosaur skeleton being reconstructed. It is a *Tyrannosaurus rex* that Disney purchased for more than seven million dollars and is being assembled by paleontologists from the Field Museum in Chicago. Disney stresses that every dinosaur footprint and faux fossil in DinoLand is archaeologically correct. The place is likely to be more memorable, however, for its play areas, chutes, climbing nets, giant sandbox, and most especially its time-machine thrill ride Countdown to Extinction, or CTX, the loudest and most raucous at any Disney park. Enormous investment in veracity is mingled here with Hollywood simplistics. Impressively realistic dinosaurs have been created as animatronic wonders, from a thirty-three-foot-long *Carnotaurus* rampaging through the primeval forest to a chicken-sized *Compsognathus* screeching as it leaps through the air. But the inevitable hero and villain are here, too—the vegetarian *Iguanadon* is given an especially benign face, looking wise and gentle, in contrast to the savagely toothed, horned, and exaggeratedly bumpy-skinned *Carnotaurus*, the "meat bull."

The same potential conflicts mingle at Harambe, the hub of Disney's Animal Kingdom. This mythical town (the name means "Come together," in Swahili) is an island of tropical greenery and equatorial architecture inspired by the African coastal towns of Lamu, Shela, and old Mombassa. Not specifically modeled on any one identifiable street or marketplace, it is an amalgam that looks like an ancient town built around the tourist economy of an African safari. It resembles traditional native construction of a hundred years ago, nonetheless designed to meet modern building codes. New walls look old, apparently weathered and stained by sand and rain storms, though only as convincingly detailed as a stage set. Corrugated metal and thatched roofs, built by Zulu craftsmen, give a vernacular casualness to the place, but are intended to withstand at least sixty years of Florida's rains and heat. None of this, though, can disguise the fact that these impressively realistic arcades, shuffling along the bustling main street, are principally intended to sell cinnamon muffins and coffee cake, exorbitantly priced drinks, baseball caps, T-shirts, mouse-eared pith helmets, shell necklaces, watches, stuffed monkeys, and bags to put all the stuff in. The place is dedicated to overconsumption and strongly contradicts any attempts to present messages of conservation and wise stewardship of the planet's resources.

The duality of the messages at Disney's Animal Kingdom is omnipresent. At one edge of Harambe is the Caravan Stage Show, where Steve Martin (the bird trainer, not the comedian) hosts a politically correct but philosophically flawed bird show, called Flights of Wonder. It is repeatedly stressed that everything the birds do here is part of their natural behavioral repertoire. Groucho, a yellow-naped Amazon parrot, whistles "Yankee Doodle Dandy" to rapturous applause.

A BIT OF AFRICA IN FLORIDA

At the other end of Harambe's main street, a gnarled, ancient baobab tree, the quintessential symbol of the African savanna, is a beacon to another type of experience, a journey through the most spectacular of this park's features.

It is here (in contrast to the ordinariness of the zoo exhibits at the Oasis or the Tree of Life), while experiencing the Kilimanjaro Safari, that one discovers the best evidence for a zoo design philosophy espoused by Disney's chief veterinarian, Peregrine Wolff: "We want to set new standards. To say that not only is it not okay to keep animals in concrete cages, it's also not okay to keep animals in beautiful natural habitats that are too small" (Malmberg, 1998).

This 120-acre African savanna, larger than San Diego Zoo's site or the entire Magic Kingdom, is the biggest feature of Disney's Animal Kingdom and is skillfully composed to appear even larger. But its size is only the first of its superlatives. Brimming with drama, it progressively unfolds glorious and very carefully controlled views that are astoundingly realistic. Kilimanjaro Safari is Disney's most ambitious virtual reality adventure in a long history of fooling all of the people some of the time. It is as close to the real savanna as one can experience outside Africa, except that here the gazelles do not get eaten by lions and the elephants are not killed by poachers.

These important variances, however, are highlighted. Visitors tour most of the savanna by a sort of jeep-lorry, an open-air camouflage-painted vehicle artfully kitted out with all the trappings of safari survival gear, with deliberately squeaky springs and softly lurching suspension, that carries thirty-two passengers along a bumpy safari track. This road has been as closely considered as almost every other detail in this exhibition habitat, to create an accurate simulation of the real thing. Made of concrete, it is colored to match the surrounding soils. Tires rolled through it and stones, dirt, and twigs dashed into it before the concrete set make it look as if it has been traveled and rutted by a thousand lorries. At one point the road is made to appear quite steep, at another it scrapes through kopje outcrops or fords shallow rivers. The route crosses a rustic bridge, skillfully engineered to be rickety, but as safe as any lawyer could wish for. It lurches at one point and the lorry hurries away, after which the bridge hydraulically corrects itself, ready to cave in beneath the next vehicle coming round the bend. Traversing a series of orchestrated landscape experiences, the journey encompasses many different types of savanna habitat. There are dense scrub areas, where the road pokes through heavy vegetation, sightlines controlled by escarpments and other geological features, and open views of stunning beauty.

The driver points out and explains features along the journey and gives out conservation information, such as the stark fact that in the past twenty years, we have been losing elephants at the rate of 150 a day. But the narrative is interspersed with obviously false admonitions ("Everybody keep their fingers crossed. This is a very old bridge we have to cross here!") and runs the risk again of the realities of life losing their probity as they rub up against the showmanship of Disney. It is the same dilemma one always faces when becoming absorbed in Disney's stories, as confusing as the false scenes of lemmings committing suicide by the hundreds in one of the earliest nature documentaries produced by the studio. It comes ominously close to implying that Nature isn't quite worthy of sustaining its own stories without embellishment. It is the perpetual impulse of the tale-teller to overstep the truth.

Even so, the beginning of the safari story is almost faultless. It begins with a passage through an area of heavy vegetation, like the preamble to a symphony, then, Beethoven-like, triumphantly emerges into an open area and a breathtaking view of the savanna, with herds of animals ranging the lush grasslands. Then come the details, the motifs among the melody. A family of mandrills lounges on a rocky area beside the road. Emerging from a jumble of granitic outcrops called kopjes, views open onto a mud wallow and white rhinos. A variety of gazelles, wildebeest, zebras, and giraffes are grazing or browsing at astonishingly close range. The impression is that one could reach out and touch them. Another kopje is the home for a pride of lions, while warthogs burrow around the edge of a geothermal pool with bubbling geysers and mudpots. Around a bend, a herd of elephants are foraging amidst trees, bathing in a river, rolling in sunset-red dirt. Flamingos grace the edges of a pool as the vehicle splashes and bucks through the waters. Prehistoric paintings adorn the rocks of an ancient kopje, but there isn't time to peruse them: the safari resolutely progresses forward. Indeed, one wishes that for a while not much would happen, so that there could be a better sense of lucky discovery after a period of quiet rumination. It is a problem endemic in zoos: how to give visitors the chance for tranquil exploration. And it is intensified here, with the incessant movement and chatter. The basic problem with this Disney experience is that it is essentially just another ride, as opposed to a safari on which one can discover and learn at one's own pace.

At about this point on the journey, the driver turns up the radio, which has been playing Swahili songs in the background. It suddenly crackles with a message from a spotter plane overhead saying that a young elephant has become separated from its mother, and poachers have been spotted. "Who's willing to help us catch the poachers?" asks the driver, as if there was any option. A "chase" ensues, with the sounds of machine-gun fire in the air. The lorry splashes noisily through a pool and between a field of exploding geysers, through dense stands of giant bamboo, palms, and big-leaf trees, finally arriving at the poachers' camp where guards have captured the bad guys, of course, and are holding them at bay with rifles. "Good job, fellas," says the driver. It is all very Disney, ending on a predictably victorious note, and it is admittedly lame drama, though one hopes that it might serve to bring home to some the awful reality of what is happening to wildlife in many parts of Africa.

An Asia section is made up of an indiscriminate collage of the architecture and jungle landscapes of India, Nepal, Thailand, and Indonesia. Temple ruins and simulated rain forest habitats provide the backdrop for a lurching white-water river raft ride. Twelve-person rafts careen past big waterfalls and bamboo thickets until, suddenly, the air is suffused with the acrid smell of smoke, and a twist in the river puts visitors face to face with a denuded slope of logged-out forest. The bare soil has been torn up by logging trucks and is rapidly eroding. Stumps and slash piles have been left burning and smoldering. It is a dramatic way to bring a conservation message home to visitors. Especially if their home is Asia. One wonders whether Disney will someday dare to show similar scenes of destruction of North American habitats. The Pacific Northwest has equal opportunities for presenting white-water trips and clear-cut ancient forests.

The Asia section also includes foot trails, like the Maharaja Jungle Trek with caves, crumbling arches, and footbridges, eroded parapets and old bridges as viewing opportunities to see animals such as Komodo dragons, fruit bats, and Malayan tapirs. Bengal tigers lounge by a pool in the overgrown gardens of a palace. Gibbons swing on bamboo scaffold around a ruin. It has a surreal sense, as if we have entered a compound where all the humans have mysteriously died and the area has been taken over by wild animals.

Another series of foot trails wander through the Africa section. These are disguised to look worn and muddy, with cracks, streaks of guano and patches of moss carefully applied with paintbrushes. The Forest Exploration Trail leads to a scientist's field house, full of nests, bones, notebooks, field photographs, computer equipment, and a colony of naked mole-rats, which are the latest hot item in zoo exhibits and here are the supposed latest project of the supposed resident, Dr. Kululunda. The back door of the house leads into an aviary, cleverly avoiding the sense of traversing a flight-lock of double doors or of entering an artificial structure. Here again Disney has been unable to resist the lure of focusing only on weird and cute species: hammerkops, jacanas, carmine bee-eaters, and pygmy geese the size of fat robins.

The trail winds out into a campsite alongside a river, where hippos can be viewed underwater, balletic à la Fantasia, in the world's biggest hippopotamus exhibit. At Gorilla Falls, two troops of gorillas are displayed in a one-acre exhibition habitat which, like so many other aspects of this development, is bigger than anything achieved at any other zoo. In one sense it represents the landscape-immersion design concept brought to its full potential; the culmination of design techniques that have been experimented and tinkered with in several zoos over the past twenty years, but here married to an awesome budget. In another way, however, it is frustratingly disappointing. The basis of landscape immersion was to give people a sense of exploring a wild habitat and of literally being submerged in a wilderness. At Disney's Animal Kingdom, there never seems to be a moment's respite from the constant noise and jostling of thousands of other people. The purpose is defeated, swamped by its commercial success.

A small army of Imagineers is credited with bringing this project to reality. Disney's promotional materials never mention, however, that Jones & Jones designed the savanna. It must have been immensely satisfying for that design firm to at last have a contract where they could engage their zoo design philosophy without the constraints of a typical city-zoo budget.

The savanna contains hundreds of species of plants, including about one hundred trees and shrubs that have never before been grown in North America. To maintain this rich landscape, horticulturists enter the savanna in the early hours of each morning and plant new trees, shrubs, and grasses for the animals to feed on each day. Water drippers for birds are hidden inside

In the twenty-minute Kilimanjaro Safari ride, visitors to Disney's Animal Kingdom get the chance to see hundreds of different animals from the African savannas. The experiences are sometimes wonderfully real, yet at the same time one feels vaguely distant from it all, due to the transitory nature and fleeting imagery of the ride. Disney has a tradition of reducing reality to simplistic stereotypes, and there is a disturbing and dangerous proximity here to a hermetically packaged vision of Africa. Even so, no other zoo has yet managed to capture such high levels of realism as in this replicated savanna. (© Disney Enterprises, Inc.)

artificial termite mounds. Feeding troughs are also disguised, inside artificial tree stumps that, at least at this zoo, look real, or, in a couple of instances, inside artificial animal "carcasses," providing, as Malmberg describes it, "an authentic-looking tale of the circle of life, without putting any of the park's animals in harm's way."

COMMERCIAL OR CONSERVATION CENTERS?

Visitors traveling through this savanna have the chance to see about one thousand animals, including elephants, lions, giraffes, cheetahs, rhinos, bongo, duiker, crocodiles, wildebeest, okapi, hippos, anteaters, zebras, ostriches, Thompson's gazelles, baboons, flamingos, cranes, marabou storks, scimitar-horned oryx, warthogs, nyala, and another dozen species of antelopes, all in natural-sized families and herds. Probably a greater variety of animals, on a twenty-minute ride, than they would have seen if they had flown to Africa and taken a real safari. That's the Disney magic. But there, too, perhaps is the Hollywood rub. Is it all a bit too easy? A bit too fantastic? Can thrill rides comfortably fit with wildlife-viewing experiences? Will Disney's grand adventure "impart a greater appreciation of the natural world" or will its various scenarios combine to form a jumbled image of parades, shopping, musical shows, shopping, rides, shopping, eating, shopping, flitting past geysers and through waterfalls, and seeing wild animals in remnants of wild places? The hazard is that, like Planet Ocean, it could reduce animals to a mere backdrop of a set of multisensory staged experiences. It is the danger that many other zoos have flirted with, in a minor key. Most visits to most zoos throughout history have served only as diversions for the curious. Most zoo animals have traditionally been reduced to caricatures of their wild cousins.

Zoos now face a challenging future. There is no doubt that they will have to expand upon or find new sources for their income. A peek into the gift shop has always been a reliable indicator of the overall quality, philosophy, and ethics of a zoological park. In the future this commercial window into their soul will become ever more telling. Will zoos, as some of their operators wish, subsume themselves entirely in commercialism? Will they follow the path that leads to emphasizing only such satisfyingly quantitative methods of measurement as coins put in the telescope, rides taken on the zoo train, popcorn spilled on the sidewalk, balloons floated into the trees, and revolutions of the turnstile? Or will something more fundamental, more qualitative, drive their future? Disney's Animal Kingdom is too large to serve as a role model in entirety, but zoos can choose to select almost anything they want from it. Unfortunately, the constant sound of cash registers might prove a more seductive siren for many zoos than any other message to be found there.

It is undeniable, however, that the design quality of Disney's replicated savanna, with its impressive realism, has set a benchmark for zoos. Since the

At the San Diego Wild Animal Park, a savanna habitat is represented by a severely eroded landscape featuring a tree inside a metal tube inside a concrete pipe inside a circle of jagged rocks. The value or purpose is mystifying. (Author photo.)

1980s, San Diego Wild Animal Park has occupied a position in the public mind and media of what an open-range zoo should look like, even though its planning and design standards are very inadequate. There is, for example, no sense of discovery, no thrill of wilderness for visitors touring this degraded site; the limited extent of the eroded enclosures has always been woefully apparent at all parts of the journey through the Wild Animal Park. Fences and other animal management paraphernalia are also always very visible reminders of the artificiality of the place. The natural splendor of real savannas never inspired San Diego's designers to reach for realism. Ironically, perhaps the commercial spur of Disney's savanna will encourage them to strive for better standards. In addition, let us hope that other principles become paramount, and that maintenance of the inherent dignity of wild animals will be seen as essential by all zoos. Certainly the animals' needs must be placed higher on the scale of decision making. There are encouraging signs that zoos are coming to recognize that their visitors are actually their owners and not merely a nuisance to be tolerated. And there are more indications today than ever before that the zoo horizon might be expanded to include not only under-represented species, but also the related areas of botany, geology, and paleontology.

The strong conservation ethics of the Bronx Zoo have not compromised that institution's success in any way. Noorder Dierenpark's decision to embrace difficult and complex stories has not reduced its economic viability. The Arizona-Sonora Desert Museum's dedication to the local ecology has not prevented it from producing surpluses at the end of the financial year (this is particularly impressive considering that all school visits to the Arizona-Sonora Desert Museum are free of charge). Nor has the commitment of each of these institutions to education and to excellence hampered their ability to operate efficiently and to be financially sound. A wise direction for the future of zoos is only a matter of choice; it is merely uncertain as to which choices will be made.

CHAPTER NINE

EPILOGUE

Humans once thought that this globe was safely placed at the center of a perfect universe. Nicolaus Copernicus, in 1543, was the first to reveal Earth's status as a tiny satellite circling one of countless stars. Although the peripheral location is no longer disputed, the central importance of humans in the universe, with a preordained destiny as beings made in the image of their creator, is a concept that many people remain reluctant to abandon.

The telescopes and microscopes of science have refocused the myopic view of our uniqueness and thereby earned special resentment from many. They are threatened by science, observes Duke University's Matt Cartmill (1993), because it "undercuts the special status we accord to human beings." We now know that more than 98 percent of human DNA is shared with chimpanzees and at least one human gene is identical to that in yeast (Jones, 1996). Some people find such knowledge even more disturbing than the existence of black holes in space, more disconcerting than the Big Bang theory. Although there should be deep comfort in recognizing that humans are inextricably linked to all living things, Darwin's theory of evolution has been subject to special vitriol. In the Netherlands in the 1990s, the education ministry repeatedly cut evolution from school examination topics recommended by scientific advisers, on the basis that it is "only a theory." Britain's

Creation Science Movement, which has some two thousand members, opened a public exhibition in February 2000, in Portsmouth, which asked such questions as "If life could arise on its own in a primeval soup, why don't we find bacteria when we open a tin of soup?" In the Australian state of Queensland, students are taught "creation science." And the Kansas Board of Education in 1999 banned teaching of the theory of evolution in state schools (MacKenzie, 2000). It is not uncommon to find people who refute the geologists' evidence that time on earth is measured in billions of years rather than the biblical few thousand or who choose to ignore the paleontologists' findings that life has not been dominated by humans since Day 6. The Creation Science Movement publishes a pamphlet, "Anglo-Saxon Dinosaurs," which cites Beowulf as proof that dinosaurs coexisted with ancient Britons.

In past days it was not unknown for those who questioned the centrality of the status of humans in time and space to be tortured and killed. There is no guarantee that such attitudes will never return; distaste for science is abundant. It is not just because science makes the fabulous prosaic or because scientists trawl the unfathomable and then so often present us with the emotional equivalent of dead fish that there is widespread resentment or that scientists are maligned figures, the idiot-savants of popular culture. Even Aldo Leopold, likening Nature to a song, saw the scientist as the sort of person who, upon hearing the song, "selects one instrument and spends his life taking it apart" with no understanding of the music (Leopold, 1966). Cartmill notes (1993) that "dread and willful ignorance of scientific knowledge has been current among intellectuals for two hundred years." British intellectual and converted rake Malcolm Muggeridge believed that science could explain nothing that really matters. In the story of his "conversion" (1988), he dispels Darwin's theory of evolution as "a far-fetched exercise in theorizing and credulity that may amuse posterity." Muggeridge claimed that "a little science takes you away from religion, a lot brings you back again." As a late convert to Catholicism, he might have merely been mindful of the shortcomings of his own degenerate past, but Muggeridge's brand of escapism reflects a surprisingly common belief that scientists have unwittingly led us into a gloomy world of shifting uncertainties. The fact that modern science has removed much of the mystery of the universe but none of the fear of the unknown has resulted in a ubiquitous and pervasive tabloid mentality that prefers fluffy mysticism over critical cynicism, believes in angels more than genetics, and chooses to put more faith in astrology than dollars into astronomy. For many, science has shed too harsh a light on our self image, "reducing our own species," as C. S. Lewis feared, "to the level of mere Nature." Humanity, as T. S. Eliot remarked, cannot bear very much reality, which may explain why millions in our society accord UFOs, telepathy, clairvoyance, and palmistry a status that is denied to Darwin.

The popular view of Darwinism as survival of the fittest as some sort of natural justification for nasty behavior in the classroom or the board room is a distortion of his words. And the common assumption that there is a natural progression through lower forms up to the perfection of human beings at the apex of evolution is entirely wrong. In addition, many people believe that humans did not evolve through natural selection, like every other living thing, but were created from a singular recipe. Part of the resistance to Darwin's theory derives from our inability to comprehend the passage of eons of time in which natural selection operates. Carl Sagan (1985) wondered what 70 million years could mean to beings who live only one-millionth as long. He likened us to butterflies who flutter for a day and think it is forever.

Homo sapiens have existed for but a brief moment of geologic time. If the length of your outstretched arm represented the span of years since life began on Earth, beginning at your shoulder, the era of the dinosaurs would cover your wrist bones. If you were to delicately trim the nail of your middle finger, you would remove all trace of human existence. Many natural history museums present such time-lines on a larger scale by depicting them on the floor so that people can walk along the illustration. This representation fits well with our view of time as something progressive, marching inexorably from simple-celled creatures through ever greater complexity toward the moment when humans appear as the inevitable, final flowering of creation.

Buddhists and Hindus by contrast see time as something cyclical, based on repetition rather than change. The Greeks, too, conceived of time as a circular phenomenon. But Christianity developed a different perspective, regarding time as something linear, taking us to some new place ahead. In 400 St. Augustine proposed the idea of one-way time: the Church could then not only determine the moment of the beginning of the world but, more importantly, prophesy its end. Two new attitudes grew from the concept of one-way time. First, that humans determine their own future and thus in some special way transcend the natural world. This in turn created an attitude that people could legitimately change the environment to suit their own purposes and justify those changes as a manifestation of divine purpose.

Evolution suggests that a horticultural analogy would be more appropriate, with a radiating pattern of multiple branches to describe the story of life on Earth rather than a simple unidirectional model. Life has never unfolded in a linear development; it is not like a train on a predetermined track. Humans are, in the words of Harvard paleontologist Stephen Jay Gould (1996), "a tiny twig on the floridly arborescent bush of life . . . that would never produce the same set of branches if regrown from seed." There is no logical progress to evolution or indeed to the eventual appearance of humans in the world. Our supposed universal essentiality is no more than collective deception, a shared egomania that blinds us to observing more useful truths.

In the random variation of natural selection, individuals that are better able to survive changes in their local environment will be more likely to reproduce, and their offspring will likely carry the genes that made that particular adaptability possible. Progress is thus not predictable, nor inevitable. Nor is complexity preordained. So-called simple creatures have survived successfully for many more millions of years than humans have. In recent times it has become clear that there has been a pattern of unrelenting emergence and extinction of life forms since life began.

THE DISCOVERY OF EXTINCTION

Only about two hundred years ago, a discovery was made that shattered the beliefs of those who were convinced the world was perfectly formed and immutable. The bones of mammoths and mastodons came to be recognized as the remnants of extinct species. It was a frightening message. Was not extinction sure evidence of cosmic failure, an attack upon the sacred principle of an unalterable Creation? Thomas Jefferson was among those who found it difficult to accept such an idea. One of his ambitions for Lewis and

Clark was that they would discover a gigantic American lion, whose existence was supposed from Jefferson's possession of the claw of an extinct giant ground sloth. Mysterious huge bones that had found their way to the White House from lands west of the Mississippi were deemed to be similar evidence of unknown living monsters in wilderness areas unexplored by Europeans.

As more proof of extinction was unearthed, and, moreover, the vast scale of this catastrophe emerged, nineteenth-century natural historians (who had still not yet learned of the past existence of dinosaurs) struggled mightily to come to terms with and to explain the causes for this phenomenal loss of so many species of large animals. When they discovered evidence of past ice ages it was supposed that violent climatic changes had been the culprit (Grayson, 1984). It would not seem surprising if there had been a huge loss of animal species under the conditions of an ice age. During the last one, which lasted maybe fifty thousand years, average world temperatures fell by almost fifty degrees Fahrenheit. The polar ice cap extended as far south as Wales and New Jersey. Europe was a polar desert. So much of the world's water was locked up in ice, the sea level fell by more than five hundred feet. Glaciers formed on mountains in the tropics (a few of these still exist, although global warming is now melting them).

Several of the previous ice ages in the last 1 million years had been even more severe, yet there is no evidence that any of them resulted in any dramatic loss of species. It seems that the animals adapted and in some way coped with the conditions. Further, whereas the formation and duration of the last ice age was global, occurring everywhere at the same time, extinction of many of the world's megafauna is staggered over fairly long periods of time. Large marsupials and giant birds disappeared from Australia at least thirty-five thousand years ago. By contrast, New Zealand's moas, flightless birds, some as tall as an elephant, became extinct just a few centuries ago (Diamond, 1984).

Paleontologist Paul Martin, of the University of Arizona, has assembled weighty and compelling arguments to explain what might have killed off so many of the big animal species that once populated the planet: he conjectures that humans caused the extinctions. Now, only in parts of Africa can anything resembling the opulent megafauna that once characterized the planet be found. There is strong evidence that humans decimated the megafauna wherever they went, clubbing the animals out of existence with an aggressive adroitness that presaged the devastation *Homo sapiens* are accomplishing in modern times.

Recognition that our cruelty to and wanton killing of animals also debases us is quite a recent revelation. Nor are we as well advanced in this lesson as we could be, though in the Western world there is more consideration today for the well-being of animals than ever before. Our concerns, however, are not always rational, and we should be surprised if they were, for animal welfare springs more from the heart than the mind. More regard is often given to an individual, for example, than to an entire species. We find it easier to care for other living things when they have been personalized and, preferably, named.

Even so, compassion for other living things does not seem to come naturally. People have to learn it. Australian research scientist Tim Flannery (1998) lived among tribes in New Guinea that had known only Stone Age culture until about fifty years ago. They had been separated from the rest of the world for tens of thousands of years until gold prospectors, in 1938, stumbled upon an unknown population of well over a million people in the New Guinea highlands. Flannery says that he often mused on the differing concepts of animal cruelties possessed by these tribal people and Westerners. The horrors that he observed, one should be warned, are very upsetting. Hunters commonly carried living possums and wallabies with all limbs broken, intestines and eyes hanging out. They would not put the animals out of their misery because keeping them alive prevented the flesh from rotting. Captured birds were routinely plucked alive and boiled alive. Pigs were blinded with lime, so they would not wander. These practices probably typify the human attitudes toward animals in the harsh world of the late Pleistocene era. We might seek to explain such actions as brutal—or "inhuman"—but they are essentially and purely human. And if we find them appalling, we need to explain how we tolerate equal atrocities in the laboratories and factory farms of our own society, choosing to remain quiet and calm about them simply because they do not happen before our eyes.

Flannery also describes another persistent human trait among the tribal peoples of New Guinea; the constant damage and mindless destruction to the environment. Young men would fell a gigantic tree, encouraging its fall with excited whoops, simply to get a strip of bark to roof a makeshift shelter. They would then fell other ancient trees to create a clearing for one season's crop, before moving on, leaving devastation behind them.

Essentially, the Irian Jayan farmers who blind their pigs are no different from the modern scientists who test cosmetics by blinding rabbits. The naked, feather-bedecked Goilala tribesmen of New Guinea who so carelessly waste big trees are in truth indistinguishable from those attired in business suits who work for the corporations that are clear-cutting the world's forests or over-fishing the seas. But neither are the artists who so skillfully painted the cave walls at Lascaux one whit different from those who now create 3D images for HDTV. Our minds and our bodies have not changed at all in at least the past forty thousand years. It is our knowledge that has expanded, not our brainpower. We cannot think any better, probably do not hear as well, and our eyes are no more advanced, though in recent centuries we have started to perceive some things more clearly.

The very properties that have ensured our evolutionary success and our worldwide distribution seem to be the same abilities that lead us to cause such tribulations for other living things: our dexterity, inventiveness, curiosity, cleverness, and aggression. We also have an innate tendency to manipulate things. Unfortunately, this aspect of human nature is expressed as much by pulling wings off flies as by building airplanes, by carving initials on a tree or tagging public property as by decoding the Rosetta Stone. The skills in butchering a mammoth with a stone tool are directly linked to our ability to be brain surgeons. Human beings or, perhaps more accurately, human *culture* has consistently if fitfully progressed toward a more refined civilization, but just as persistently has always had devastating effects upon Nature. Presently, human culture is transforming the world at a terrifying rate. Those who allegedly wiped out entire species of large animals through the late Pleistocene were no more aware of the consequences than the New Guinea tribesmen, with their belief in a never-ending supply of Nature's bounties. They were ignorant of the knowledge we now possess and that we must not continue to ignore.

Humans did not have to invent either devils or angels; they exist within us. We have the capacity to create immense delight and goodness and to reach the deepest evil. The atrocities of some individuals can astonish as much as the goodness and courage of others. The mingling in all humans of cruelty and compassion is an impossible quandary. Alexander Solzhenitsyn, after decades of suffering in Stalinist Russia, learned that the line dividing good and evil cuts through the heart of every human being (Masters, 1996). Humans, throughout history, have casually and callously decimated countless other species, yet in our future we can, if we wish to make the effort, save many more that we have endangered. Though we have always been able to kill very effectively, we are also beginning to become adept at the much more difficult task of mending things. The making of laws to govern our behavior is a mark of civilization, and we have considerable progress to make in controlling the extent of our destruction of the natural world.

Before humans migrated to North America, the continent was roamed by forty-five genera of very large animals. Today only twelve exist; some of them, like bison, only just. None of America's giant sloths, sabertooths, cheetahs, giant beavers, glyptodonts (armadillo-like creatures, the largest of which were the size of a Volkswagen), short-faced bears, two types of capybara, several species of three-toed horses, tapirs, camels, llamas, saiga antelope, yaks, or the elephantine gomphotheres, mammoths, and mastodons are extant. They disappeared after the arrival of humans, presumably hunted to extinction. South America had fifty-eight genera of megafauna before human contact but now contains fewer than a dozen. Australia may have lost at least thirteen genera of large animals after the arrival of the first humans on that island continent, which was once home to marsupial lions, giant kangaroos, diprotodonts—one like a two-ton rhino—and gigantic flightless birds. (The Anglo-Celtic races have continued the tradition. Almost 10 percent of Australia's mammals have become extinct in the past two hundred years.) It took a longer time before people started to discover the planet's smaller islands, but when they did, the same pattern of destruction was repeated. Pygmy mammoths survived on islands in the Arctic Sea and dwarf elephants (some no bigger than a very large dog) on several Mediterranean islands until human contact just four or five thousand years ago. Polynesians arrived in Hawaii about fifteen hundred years ago, and the world soon lost several species of flightless birds. The same happened throughout all the South Sea Islands; a dazzling range of previously unknown animals were soon extinct. Just one thousand years ago, New Zealand had at least twenty-five species of moas. Humans then arrived, and the last moa was killed about five hundred years later. To add further injury, destructive species like rats, cats, and mosquitoes were introduced to the scattered paradises in the South Seas.

Naturalist Alfred Russel Wallace, who had independently developed a theory of the origin of species, was also one of the first naturalist philosophers to ponder how enormous numbers of big animals had disappeared in geologically recent times. In 1876 he noted: "We live in a zoologically impoverished world, from which all the hugest, fiercest, and strangest forms have recently disappeared" (Raby, 1996). More than a hundred years passed before Paul Martin's "blitzkrieg hypothesis" gave the world a probable answer to the puzzle (Martin, 1984). All over the world there is evidence that within a few hundred years of the appearance of humans in any particular territory, the hugest animals were often exterminated. Wallace had earlier intuitively recognized the cause for this impoverishment. Marveling on the beauty of birds-of-paradise, he wrote how sad it was "that on the one hand such exquisite creatures should live out their lives and exhibit their charms in these wild inhospitable regions [but] should civilized man ever reach these distant lands, and bring moral, intellectual and physical light into the recesses of these virgin forests, we may be sure that he will so disturb the nicely-balanced relations of organic and inorganic nature as to cause the . . . extinction of these very beings whose wonderful structure and beauty he alone is fitted to appreciate and enjoy" (Wallace, 1876). Setting aside the fact that birds-of-paradise might be even better suited to appreciate the structure and beauty of their own kind, for that is why they have evolved their exuberant plumage and ecstatic dances, Wallace was nonetheless accurate in his assessment of the apparently inevitable destruction that we bring with our morality and our intellectual superiority.

HOW WE LOST THEIR TRUST

There is good reason why wild animals are so exceedingly wary of humans. Yet things were not always this way. As Tim Flannery points out, it is humans who made the savage beast. To see how wild animals behaved before they learned to distrust people we need only look to some recent examples of places uninhabited by our kind.

When Charles Darwin visited the Galapagos Islands in 1835 he noted the extreme tameness of the birds on one island unpopulated by humans. "One day . . . a mocking-thrush alighted on the edge of a pitcher . . . which I held in my hand, and began very quietly to sip the water; it allowed me to lift it from the ground while seated on the vessel." People who have the good fortune to visit Antarctica experience the same sort of delight, walking among penguins that are blissfully ignorant of our killing power. On another island in the Galapagos, which had been colonized just six years earlier, Darwin also wrote in his *Voyage of the Beagle,* that he saw a boy "sitting by a well with a switch in his hand, with which he killed the doves and finches as they came to drink." Our willingness to maim and kill is sometimes so casual.

The case of Kangaroo Island (off the coast of South Australia) documented by Flannery in his ecological history of the Australasian islands (1994) is especially revealing. When Matthew Flinders visited in 1802, the island had been uninhabited by humans for thousands of years. Flinders described the kangaroos on the Kangaroo Island as incredibly docile. His crew members killed them with sticks. The animals had been isolated from humans long enough to have lost fear of them. Another important piece of the puzzle is found in the different fates of the elephant seals and the sea lions of King Island, in the Bass Strait, also visited by Flinders in 1802 shortly before British sealers took up residence there but thousands of years after humans had last lived on the island. Elephant seals went to the island for breeding, after spending the greater part of the year feeding in the sub-Antarctic ocean. Without knowledge of humans, these enormous animals, weighing up to four tons, watched uncomprehendingly when the sealers walked among them and shot or clubbed them. Every one of them was eventually killed. Elephant seals have never returned to King Island. (The indigenous emu, wombat, and spotted-tailed quoll were equally trusting. Soon they were all dead, too.) But sea lions also lived on this island in 1802, and they are still there. Sea lions forage in coastal waters and had undoubtedly visited the mainland often enough to gain experience of humans. They were wary enough to survive when sailors landed on King and Kangaroo

Islands. And herein lies a clue as to why large animal species still exist in such numbers on the African continent.

Africa has certainly lost many species of megafauna. Thirty of its thirty-seven genera of large mammals became extinct in the late Pleistocene, including many species of primates, giant cats, mammoths, chalicotheres (like heavy-limbed horses with claws), gomphotheres (four-tusked elephants), the moose-giraffe, and several bovids. But Africa is also the only continent to retain significant numbers of late Pleistocene species. This might seem paradoxical, because Africa is where humans originated. Why were they not wiped out first of all? Tudge (1995) conjectures that there are probably three reasons. Parts of Africa have apparently always been cauldrons of infection—malaria, yellow fever, trypanosomiasis, and ebola, for examples—and humans have been unable to populate these areas in significant numbers. Also, the large animals that survive in Africa tend to be either solitary and secretive forest dwellers or migratory savanna species which, without the horse, could not be effectively hunted. Moreover, and perhaps especially important, African megafauna and humans were evolving together for about a hundred thousand years. The animals had been learning to avoid flying sticks and stones since human hunters first started trying to kill them. The people who walked into America and Australia, however, were skilled and proven hunters, pitted against naive prey.

Humans have been butchering wild animals relentlessly since then. Quite recently people have also become efficient global destroyers of their habitats. In 1945 Fairfield Osborn addressed a meeting of the New York Zoological Society with prescient vision: "We need to face the unfortunate fact that there are two world wars going on. One is man's destruction of man; the other his destruction of the living resources of nature" (Bridges, 1974). Looking at photos of World War II seems to give a glimpse into another age. But the other world war that Osborn referred to continues yet and has dreadfully intensified. Since the advent of the Industrial Revolution, humans have been involved in such destruction that one might think we were not only at war with Nature but wanted to see it annihilated. Europeans have waged a battle against their forests equivalent to a scorched-earth policy, as evidenced by the fact that "Holland" once meant "Land of the Forests."

Overhunting since the time of the last ice age may have caused the decimation of much of the planet's megafauna. Their extinction was probably not intentional but merely the product of human short-term gain. The Mexican truck driver who a few years ago shot the world's last imperial woodpecker, the largest of the species, commented only that it tasted good. Being insatiably greedy, we now consume 40 percent of the terrestrial primary productivity of the planet (Conway, 1999). In the same way that early hunters assumed that their prey was limitless and that the world stretched forever beyond their horizon, people today seem to think that they can drain wetlands for shopping malls, cover fertile valleys with factories, and take whatever they want from Nature with impunity. Forests on the hillsides are clear-cut with no recognition that this causes water tables in the lowlands to collapse. Plants and animals are being exterminated at an unprecedented rate, with no recorded knowledge of their potential medicine or food values and no sense of their place in ecology, and frequently being sent unnamed into the abyss of extinction.

THE SIXTH IMPLOSION

Yet extinction is also a natural phenomenon. Since life evolved on Earth, species have been falling into extinction at an average rate of about one every million years (Ehrlich and Ehrlich, 1981). However, there have been at least five great implosions of extinction. The Ordovician extinction crisis occurred 439 million years ago; the only animal life on the planet at that time was in the seas, and about 85 percent of all species disappeared. Another extinction crisis of similar magnitude, the Devonian, hit the earth 367 million years ago. The worst of the five mass extinctions was the Permian, about 245 million years ago, when almost the entire animal kingdom of the time disappeared, with an extinction rate of 95 percent. The Triassic extinction crisis saw another massive drop 208 million years ago, and the most recent, the Cretaceous extinction, saw the loss of more than 75 percent of all animal species, including the dinosaurs. After each of these cataclysmic events, Nature recovered its biological diversity. But it takes many millions of years for this to happen. Meanwhile a sixth human-caused massive decline is now underway, and in less than one century humans have the capability and are proceeding to reduce the diversity of species to such an extent that it could take 10 million years for evolutionary recovery (Kirchner and Weil, 2000). In particular, it should be noted that humans are eradicating not just species but what E. O. Wilson (1992) calls the Theaters of Evolution; those natural environments in which evolution can begin to re-create biological diversity.

Since 1950 about one-third of the world's forests have been cleared and not replaced; one-fifth of the world's topsoil has been exposed and then washed or blown away and continues to be lost at an estimated twenty-four billion tons a year, equivalent to all the topsoil on all the wheatlands of Australia; carbon dioxide has increased by about one-third; stratospheric ozone is disappearing; and biological diversity is diminishing hundreds of times more rapidly than at any time during the past 65 million years.

At the same time we are despoiling our own built environment. It is said that the people shape the buildings, and the buildings shape the people. Of even more import is that human cultures are shaped by their landscape. For increasing numbers of people that landscape is a world of steel, glass, and concrete; an environment of high-rise towers, ricocheting noise, internal-combustion pollution, roads, sidewalks, gutters, and walls. It is a world of flat surfaces and geometrical shapes. To complain might seem ridiculous, for humans built this world consciously and deliberately. We develop universities in which student planners and engineers and architects learn how to create this type of environment.

There are many reasons, of course, to be grateful for a domestic world. There are untold advantages to living in a heated house in winter with a supermarket, laundry, and dentist down the road and emergency services in reach by telephone. It was not many generations ago (for most of us) that our ancestors lived in huts of sticks or mud, with insufficient ventilation and light, existing on a diet of meager sustenance and unvarying monotony, constantly praying for more rain, or less rain, suffering toothache and lice, knowing that a broken leg could mean death, ever fearful of smallpox, pregnancy, and marauding neighbors.

But the world of those recent ancestors also put them into daily contact with bird song, and starry skies, and long walks over the hills. They had a detailed consciousness of the way things grow. It was a world where people took fruit directly from trees and edible roots from the ground and fish from wild streams. Novelist Julian Barnes (1999), writing of an England now past, observes that once "the progress of winter was calibrated by the decay of racked apples and the increasing audacity of predators," whereas in the modern world the weather has "been diminished to a mere determinant of personal mood." To be aware of these fundamental changes is not just a romantic sense of loss. As modern society has distanced itself from the natural world, people have deluded themselves that they are not, now, a part of Nature.

LOSING CONTACT

Gary Nabhan and Sara St. Antoine (1993) recently surveyed children living within a twenty-five mile radius of two national parks—Organ Pipe Cactus National Monument, on the Arizona border with Sonora, and Saguaro National Park, west of Tucson—in one of the most sparsely populated areas of the United States. The children came from American Indian, Mexican Indian, Anglo, and Hispanic backgrounds, and from a range of socio-economic levels. The survey revealed disturbing trends. Most of the kids had gained their knowledge of the natural world not through multisensory experiences in the richly textured landscapes around them, but from the two-dimensional world of television and books. They may have garnered more facts about exotic species than their grandparents ever knew, but few of them were aware that the local desert birds sing more in the early morning than in the middle of the day. None of the Anglos and only about half of the Hispanic children knew that the fruit of the prickly pear was edible, yet this fruit has been a major food source in the area for at least eight thousand years. Most worrying, a clear majority in each category had never spent more than half an hour alone in a wild landscape.

While our knowledge of the details of wild animals, fed by beautiful books and even more gorgeous television documentaries, has never been greater, our understanding of Nature now has a precariously weak tenure. As Nabhan and St. Antoine point out, replacing direct experience with television-mediated education discourages children from making their own observations and forming their own opinions of the natural world. Many television documentaries inevitably focus on the spectacular, whereas half an hour alone in a wild landscape trying to get a face-to-face encounter with a tiny lizard might be even more instructive than seeing a Komodo dragon on the TV screen.

A spate of recent projects in Europe and Japan reveals our peculiar appetite for sanitized landscapes. Hermetically sealed domes have been built that contain replications of seasides, with imported and pure white sand beaches, bays of chlorinated water, concrete cliffs, and artificial surf, beneath steel-latticed roofs. There are no tidal rock pools for kids to explore, no mud to squish between toes, no seaweeds or smooth stones to collect, no shells to take home. Of course, there are no sand midges, either, and no chance of sunburn. Instead, a homogenized version of Nature guarantees permanently blue skies and consistently warm temperatures, together with abundant opportunities to eat and drink. One of the largest of these projects, the Ocean Dome at the most inappropriately named SeaGaia, a resort development on Kyushu Island, Japan, also provides a volcano that regularly rumbles and sprouts flame. To create this technological seaside, SeaGaia destroyed hundreds of acres of what had once been ecologically rich natural landscape, spending $2 billion to replace it with the usual mix of tennis courts, a golf course, hotels, and restaurants, plus the Ocean Dome and artificial bays for "safe" sailing. The corporation behind the development claims that the sea is dangerous for windsurfing and polluted so they are performing a social service by "solving such problems."

As Nature has receded from our daily contact it has slipped from view of many people, yet, perversely, acquired almost religious qualities for some. Olaf Skarsholt (Clinebell, 1996) once pictured our planet as a globe just a few feet in diameter, floating above a field, and wondered what it would then mean to us. He conjectured that "people would come from everywhere to marvel at it. They would walk around it, marveling, . . . [and] declare it as sacred, because it was the only one, and they would protect it. . . ." That image has materialized. A generation has grown up with a new view of the earth. Posters of our blue sphere hanging in space are ubiquitous in kindergartens. We have come to see, as surely as we once knew the rising and setting of the sun, that our world is a small and unique marble spinning in empty space. Astronaut Ulf Merbold first saw Earth from space and said he was "terrified by its fragile appearance." Having learned to build spacecraft that allow us a new view of the planet, we now have to learn how to live on it with much more consideration for its health.

Our divorce from the natural world is accompanied by a realization of the inadequacies of our built environments. The Age of Enlightenment's dream of mastery over Nature has turned in on itself. People once assumed that technologies would increase our leisure and beautify the world. Instead they have led to a culture of workaholism and to ecological massacre. The seductive promises of the modern world create despair for those locked in the urban ghettos and primitive villages who cannot join it, yet bring no lasting satisfaction to those who worship at its commercial altar.

Isaiah Berlin, philosopher and champion of cultural pluralism, admitted that he had "no idea how one stops one group, one race, from hating another." A potential answer lies in encouraging a universal love of Nature, with its subsequent recognition that we all share the only planet we know that harbors life. One touch of Nature, said Shakespeare, makes the whole world kin. If we restrict our philosophic wanderings to only the mind of humankind the answers to human problems must remain elusive. Nature reveals all the truths we need, showing us that we do not require our invented and dangerous concepts of deities, nations, and races. None of these things exist other than in our febrile imaginations. If we looked instead to Nature for guidance, we would find that our salvation lies in cooperation and in sharing the world with all other life forms. We would learn to see the wisdom in Henry Beston's admonition: "Do no dishonor to the Earth lest you dishonor the spirit of man" (Beston, 1928).

It is supremely ironic that the human progress that has caused such environmental degradation and loss of life forms has also generated a new compassion for wilderness. Having won a war of depredation against wild animals, humans have become concerned for preserving those that remain. Today we have an edenic view of tropical rain forests, and they have become the icon of sacred wild lands. Now that wild places are ever more confined to isolated spots of the globe, locked up as national parks, fragmented by farms, suburbs, and commercial developments, shrinking before our eyes, people are coming to regard them with a sentimental sympathy. Controlled, subdued, and safe, wilderness is at last worthy of our benign attention.

Ever since the advent of the agricultural revolution, and until modern times, "wilderness" was also "wasteland." Uncultivated lands came to be seen as threatening places inhabited by big and dangerous wild animals and evil spirits. The wilderness was a moral vacuum where mystics sought visions and tests of endurance, running the risk of meeting the devil but taking the chance of finding a deity. The boundaries between natural and supernatural are out there, in the wild places.

NATURE AS PARADISE

Only quite recently have we come to see Nature as something worthy of inspiration, imbued with spiritual values. Henry David Thoreau, exploring the wilderness of Mount Katahdin in Maine, realized, as his ancestors in the forests of Europe had found, that wild places can be forbidding and even terrifying. In his reflections on wilderness in *The Maine Woods,* he described the place as "grim and savage." And Daniel Boone, a tougher and more rugged individual, said that parts of the southern Appalachians were "so wild and horrid that it is impossible to behold them without terror" (Bryson, 1997). As fundamental a source as childhood fairy tales records that the woods were always populated with ogres and demons. Nonetheless, Thoreau in 1862 came to realize that in wilderness was the preservation of the world. And when John Muir arrived in the Sierra Nevada in 1869, he declared that "No description of heaven that I have ever heard or read seemed half so fine" (Schama, 1995). His comparison is illuminating. Christianity, an essentially agrarian-based faith, with a formal denial of the sensual aspects of human nature, has traditionally seen Nature as something to be enjoyed only in terms that illustrated religious themes. Visions of heaven are based on natural paradises, but are extremely selective. Civilized lions and lambs are allowed into paradise, but serpents are excluded. Prayers on behalf of allowing resurrection for certain dogs are not unknown. People like to think that their pets go to heaven, especially those that understood every word said to them. But if cats and dogs have access to heaven, do they also carry their fleas with them? Does a pet guinea pig receive admission, while a guinea pig dispatched in a laboratory goes to hell? And if poodles or pet parakeets are candidates for the afterlife, by what logic could the millions of cows and chickens that are slaughtered every day be excluded? Pigs are as smart as dogs and as deserving of an eternal paradise. Presumably worms, scorpions, mosquitoes, rats, cockroaches, ants, and other animals that offend us would not make it past the pearly gates. But what about gorillas, chimpanzees, and the sex-crazed bonobos, all so very like us? All these are facetious questions, of course, and serve only to try and expose the hopelessly confused dichotomies in our prejudiced views of various animals and to underscore how distorted are our values of Nature.

Zoos, unfortunately, perpetuate many of these stereotypic perspectives, promoting the notion of exclusion or acceptance based on rarity or visual appeal. Disney's Oasis exhibit is populated only by animals that are pretty, cute, not too big, too fast, or too loud, perhaps weird but definitely nonthreatening. This particular oasis, under almost perpetually blue skies, with pathways meandering through gardens of tropical palms, big-leafed shrubs, and an abundance of flowering plants, with rippling streams and waterfalls, and with cafés offering milk and honey (as well as blueberry muffins and chocolate ice cream), represents a universal concept of paradise. No unpleasant animal odor shall assault thy nose, no decaying logs or tangled vine spoil thy view, and no ugly beasts offend thine eyes. No rain shall fall on thy parade. In many important ways it is as confusing and as wrong as the distorted views of those bad and ugly zoos, with their concrete walls and rusty fences.

The manner in which zoos present animals to people directly affects popular opinions about wild creatures. Zoo displays literally shape public attitudes. Visitors often do not see zoo animals as wild animals because they are not presented as wild animals. They play with beer barrels, bowling balls, and plastic milk bottles. They feed from piles of chopped carrots, heaps of premade chows, beef patties in stainless steel dishes. When Woodland Park Zoo in the 1970s started feeding whole sheep and goat to the big cats and entire rabbits and chickens to the small cats, many concerned citizens were aghast, repelled by the sight of flesh being torn from recognizable carcasses. One wrote to the *Seattle Times* to complain, explaining that in the old days the cats used to receive only "nice chunks of meat." Usually, zoo animals are in display areas where a plethora of fences, posts, waste bins, metal gates, curbs, drains, and a host of other suburban paraphernalia destroy any sense of wildness. Amidst such muddled images, it is no wonder that people have confused attitudes. They talk to animals in zoos, for example, and quite openly. Zoo goers commonly think that the animals smile for them, display a new baby for their approval, or show off for the humans' entertainment. Visitors routinely tell the animals to pose for their cameras and seem to believe that they do.

We might be standing today on what museologist Robert Sullivan calls a cultural fault line. It was not very long ago that we were comfortably secure in the knowledge that humans were placed by divine design at the top of an ascending spiral of progress, with the white man, especially, at the apex of civilization. Now we see that we live within interconnected ecological systems and interdependent, pluralistic cultural systems. More revelations are surely yet to come, and better understandings are long overdue. Certainly we need a better sense of deep time, of how life has evolved, how evolution functions. A clearer acknowledgment of our perpetual history of killing, despoiling, and extinction is required, not to seek redemption, but to remind us to change our destructive ways. Our capacity for hatred and evil is surely controllable? Can we not govern our greediness?

There is still time to change our ways, our standards, our perspectives. We can, for example, break our filthy habit of fossil fuel consumption without having to abandon automobiles or heated homes. We do not have to pollute our potable water supplies or imperil the world's ecosystems to maintain a healthy quality of life. Michael Braungart, a German chemist, and William McDonough, an American architect, are two of the new generation of thinkers who are devising ways of productivity that regenerate rather than deplete natural resources. Braungart in 1984 devised a method of bleaching paper with oxygen rather than chlorine, thus avoiding the production of carcinogenic dioxins. In 1987 he developed the first refrigerator to run without chlorofluorocarbons. McDonough in 1985 designed the headquarters for the Environmental Defense Fund, in New York City, featuring a tree-lined boulevard inside the building, and interior offices lit by daylight. His recent office complex for the Gap Corporation includes an insulating roof made of grass and a raised floor that allows the building to be flushed and cooled with air at night.

Now in partnership, Braungart and McDonough are promoting a design philosophy that calls for industry to manufacture products and materials which, after their useful life, are not simply discarded but provide nourishment for something new. They argue for two closed-loop industrial cycles—one technical and one biological, which they refer to as metabolisms. The technical metabolism would include materials that are not biodegradable and that would be continually reused. The biological metabolism would strictly exclude mutagens, carcinogens, heavy metals, endocrine disrupters, and toxins. Its products (such as containers, furniture, carpets, clothes) could then be safely returned to the organic cycle, consumed by micro-organisms in the soil (McDonough and Braungart, 1998).

Models for this design approach surround us in Nature, where nothing is wasted. Fallen leaves, decaying fruits and blossoms, dead bodies are all useful in some way to a healthy ecosystem. We have the capacity to design industrial systems that are based on similar cycles of nutrients and metabolisms. Our natural history institutions can play a critical role in helping achieve a better understanding of how this can be done, and why we need to change the way we do things. The National Museum of Natural History, in Washington, D.C., has come to recognize, says Sullivan (1992), that if it does not produce interpretive exhibits dealing with broad ecological issues now, it will be engaged in only eulogies later. All other such museums and all of our aquariums, botanical gardens, and arboretums could usefully contribute, creating a higher degree of enlightenment. Few institutions, moreover, are better suited to engaging this effort than zoos. They have enormous capacity to make strong and meaningful contact with millions of people who visit them each year, desperately looking for connection with the world of Nature.

The twenty-first-century zoological parks, like every other natural history institution, must cooperate more effectively, broadening their activities to demonstrate the interdependency of all life forms. Fred Koontz, curator of mammals at the Bronx Zoo, has recommended that zoo accreditation programs should be revised to assess levels of involvement in conservation projects in the wild (Conway, 1995c). The ultimate goal is that zoos become energetic, passionate, and skillful protectors and advocates of all things in Nature. The new zoos must engage themselves directly in new ventures and become active champions of wildlife conservation and environmental survival and breed empathy and tolerance as much as they breed emus and tigers.

For much of their history, zoos have affirmed only an imperial mastery over Nature. When operated with intelligence and compassion, however, zoos can be a most effective conservation tool. How can they best achieve this? By becoming conservation centers, by placing Nature preservation at the center of all their efforts, by reaching for the highest standards in all their projects and activities, by seeking to awake, enthrall, and educate, by articulating the wonderful benefits of conserving biological diversity, and by making strategic alliances with other cultural and natural history institutions. Zoos have the marvelous potential to develop a concerned, aware, energized, enthusiastic, caring, and sympathetic citizenry. Zoos can encourage gentleness toward all other animals and compassion for the well-being of wild places. Zoos can cultivate environmental sensitivity among their hundreds of millions of patrons. Such a populace might then want to live more lightly on the land, be more careful about using the world's natural resources, and actually choose to vote for politicians who care about the wild inhabitants of the Earth and the health of the wild places that remain. To help save all wildlife, to work toward a healthier planet, to encourage a more sensitive populace: these are the goals for the new zoos.

BIBLIOGRAPHY

Addison, Richard A. 1933. Why rocks? *Parks and Recreation Journal* 16.

Allin, Michael. 1998. *Zirafa.* London: Review.

Arizona-Sonora Desert Museum. 1997. *Conservation mission statement and long-range strategy plan, 1997–2002.* Tucson, Ariz.

Attenborough, David. 1979. *Life on Earth.* London: Collins.

Austin, William A. 1974. *The first fifty years: An informal history of the Detroit Zoological Park.* Detroit: Detroit Zoological Society.

Australian Academy of Science. 1976. *Report of a committee on the problem of noise.* Canberra.

Baetens, Roland. 1995. *The chant of paradise.* Antwerp: Antwerp Zoo.

Barber, Richard. 1992. *Bestiary.* London: Folio Society.

Barnes, Julian. 1999. *England, England.* London: Picador.

Bartlett, Abraham D. 1900. *Life among wild beasts in the "zoo."* London: Chapman and Hall.

Beston, Henry. 1928. *The outermost house.* London: Selwyn and Blunt.

Bird, Laura. 1997. Move over mall rats. *Wall Street Journal* (8 July).

Blunt, Wilfrid. 1976. *The ark in the park: The zoo in the nineteenth century.* London: Hamish Hamilton.

Bridges, William. 1974. *Gathering of animals: An unconventional history of the New York Zoological Society.* New York: Harper and Row.

Brightwell, L. R. 1936. *The zoo you knew?* Oxford: Blackwell.

———. 1952. *The zoo story.* London: Museum Press.

Broderip, W. J. 1847. *Zoological recreations.* London: Colburn.

Brown, J. H., and E. J. Heske. 1990. Control of a desert-grassland transition by a keystone rodent guild. *Science* 250.

Bryson, Bill. 1997. *A walk in the woods.* London: Doubleday.

Buchmann, Stephen L., and Gary Paul Nabhan. 1996. *The forgotten pollinators.* Washington, D.C.: Island Press.

Burnette, Curtis. 1986. The Louisiana swamp exhibit. AAZPA Annual Conference Proceedings. Wheeling, W.Va.: AAZPA.

Cannadine, David. 1989. *The pleasures of the past.* New York: Norton.

Carcopino, Jérôme. 1991. *Daily life in ancient Rome.* London: Penguin.

Carpenter, Betsy. 1992. Upsetting the ark. *U.S. News & World Report* (24 Aug.).

Cartmill, Matt. 1993. *A view to a death in the morning: Hunting and nature through history.* Cambridge, Mass.: Harvard University Press.

Clinebell, Howard. 1996. *Ecotherapy: Healing ourselves, healing the Earth.* Minneapolis: Fortress Press.

Clough, Arthur Hugh, ed. 1992. *Plutarch's lives of the noble Grecians and Romans.* Trans. John Dryden. New York: The Modern Library.

Coe, Jon. 1986. Towards a coevolution of zoos, aquariums, and natural history museums. AAZPA Annual Conference Proceedings. Wheeling, W.Va.: AAZPA.

———. 1995. The evolution of zoo animal exhibits. In *The ark evolving,* ed. Christen M. Wemmer. Washington, D.C.: Smithsonian Institution.

Collins, N. M., and J. A. Thomas. 1991. *The conservation of insects and their habitats.* London: Academic.

Comfort, Alex. 1966. *Nature and human nature.* London: Weidenfeld and Nicolson.

Communications Consortium Media Center. 1994. *An analysis of public opinion on biodiversity and related environmental issues, 1990–1994.* New York: Consultative Group on Biological Diversity.

Conway, William. 1995a. Wild and zoo animal interactive management and habitat conservation. *Biodiversity and Conservation Journal* 4.

———. 1995b. Zoo conservation and ethical paradoxes. In *Ethics on the ark,* ed. Bryan G. Norton, Michael Hutchins, Elizabeth F. Stevens, and Terry L. Maple. Washington, D.C.: Smithsonian Institution.

———. 1995c. The conservation park. In *The ark evolving,* ed. Christen M. Wemmer. Washington, D.C.: Smithsonian Institution.

———. 1999. The changing role of zoos in the 21st century. *EAZA News* 29 (Jan.).

Corliss, Richard. 1998. Beauty and the beasts. *Time* (20 Apr.).

Correll, Terrie. 1994. *Addax international studbook.* Palm Desert, Calif.: The Living Desert.

Croke, Vicki. 1997. *The modern ark.* New York: Scribners.

Cronon, William. 1995a. Introduction: In search of Nature. In *Uncommon ground,* ed. William Cronon. New York: Norton.

———. 1995b. The trouble with wilderness. In *Uncommon ground,* ed. William Cronon. New York: Norton.

Crowcroft, Peter. 1978. *The zoo.* Milson's Point, NSW, Australia: Mathews/Hutchinson.

Curtis, Lawrence. 1968. *Zoological park fundamentals.* Washington, D.C.: National Parks and Recreation Association.

Darwin, Charles. 1845. *The voyage of the* Beagle. London: Murray.

———. 1869a. *Journal of researches.* New York: Appleton.

———. 1869b. *The origin of species.* New York: Appleton.

Davis, Susan. 1995. Touch the magic. In *Uncommon ground,* ed. William Cronon. New York: Norton.

de Courcy, Catherine. 1995. *The zoo story.* Ringwood, VIC, Australia: Penguin.

Diamond, Jared. 1984. Historic extinctions. In *Quaternary extinctions,* ed. Paul S. Martin and Richard G. Klein. Tucson: University of Arizona Press.

———. 1995. Playing god at the zoo. *Discover* (Mar.).

———. 1997. *Guns, germs and steel: A short history of everybody in the last 13,000 years.* London: Jonathon Cape.

Downey, Roger. 1980. Beyond the cliché. *Landscape Architecture* 70, 6 (Nov.).

Ehrenfeld, David. 1993. *Beginning again.* New York: Oxford University Press.

Ehrlich, Paul R., and Anne H. Ehrlich. 1981. *Extinction: The causes and consequences of the disappearence of species.* New York: Ballantine.

Ehrlinger, David. 1990. The Hagenbeck legacy. In *International zoo yearbook.* Vol. 29. London: Zoological Society of London.

———. 1993. *The Cincinnati Zoo and Botanical Garden from past to present.* Cincinnati Zoological Society.

Embury, Amanda. 1992. Gorilla rain forest at Melbourne Zoo. In *International zoo yearbook.* Vol. 31. London: Zoological Society of London.

Evans, I. O. 1933. *The world of tomorrow.* London: Archer.

Ferguson, Jackie. 1992. *Roar!* Newsletter of the Fort Worth Zoo (spring).

Finlay, Ted, Lawrence R. James, and Terry L. Maple. 1988. People's perceptions of animals: The influence of zoo environments. *Environment and Behavior* 20, 4.

Finney, Colin. 1993. *Paradise revealed.* Melbourne: Museum of Victoria.

Fisher, James. 1966. *Zoos of the world.* London: Aldus.

Flannery, Timothy Fridtjof. 1994. *The future eaters.* Chatswood, NSW, Australia: Reed.

———. 1998. *Throwim way leg.* Melbourne: Text.

Fleming, William. 1974. *Arts and ideas.* New York: Holt, Rinehart and Winston.

Flint, Richard W. 1996. American showmen and European dealers: Commerce in wild animals in nineteenth-century America. In *New worlds, new animals,* ed. R. J. Hoage and William A. Deiss. Baltimore: Johns Hopkins University Press.

Fouraker, Michael, and Torren Wagener. 1996. *American Zoo Association rhinoceros husbandry resource manual.* Fort Worth: Fort Worth Zoological Society.

Gale, Oliver M. 1975. The Cincinnati Zoo: One hundred years of trial and triumph. *Cincinnati Historical Society Bulletin* 33, 2.

Garner, Robert L. 1896. *Gorillas & chimpanzees.* London: Osgood, McIlvaine.

———. 1900. *Apes and monkeys: Their life and language.* London: Athenæum Press.

Gautier, Théophile. 1869. *Ménagerie intime.* Paris: Lemerre.

Gilbert, Bill. 1997. Who's zoo? *Sports Illustrated* (Sept.).

Goode, G. Brown. 1897. *Museums of the future.* Annual Report. Washington, D.C.: U.S. National Museum.

Gould, Stephen Jay. 1996. *Life's grandeur: The spread of excellence from Plato to Darwin.* London: Jonathon Cape.

Graetz, Michael J. 1995. *The role of architectural design in promoting the social objectives of zoos.* Master's thesis. National University of Singapore.

Grandy, John. 1992. Zoos. *HSUS News* (summer).

Grant, Edward. 1997. When did modern science begin? *The American Scholar* (winter).

Grayson, Donald K. 1984. Nineteenth-century explanations of Pleistocene extinctions. In *Quaternary extinctions,* ed. Paul S. Martin and Richard G. Klein. Tucson: University of Arizona Press.

Greene, Melissa. 1987. No rms. Jungle vu. *The Atlantic Monthly* 260, 6 (Dec.).

Groombridge, Brian, and Martin D. Jenkins. 2000. *Global biodiversity: Earth's living resources in the 21st century.* Cambridge, Eng.: World Conservation Monitoring Centre.

Guillery, Peter. 1993. *The buildings of London zoo.* London: Royal Commission on the Historical Monuments of England.

Hagenbeck, Carl. 1910. *Beasts and men.* Trans. Hugh S. R. Elliot and A. G. Thacker. London: Longmans, Green.

Halmi, Robert. 1975. *Zoos of the world.* New York: Four Winds.

Hancocks, David. 1971. *Animals and architecture.* London: Evelyn.

———. 1980. Bringing nature into the zoo. *International Journal for the Study of Animal Problems* 1 (summer).

———. 1981. Naturalistic solutions to zoo design problems. In *Third international symposium on zoo design and construction,* ed. Peter Stevens. Paignton, Eng.: Whitley Wildlife Trust.

———. 1989. Seeking to create illusions of wild places. *Landscape Australia* 2, 3 (Aug.) and 2, 4 (Nov.).

———. 1990. Seeking to create illusions of wild places. *Landscape Australia* 12, 1 (Feb.).

———. 1994. Zoological gardens, arboreta and botanical gardens: A trilogy of failure? Proceedings of the American Association of Botanical Gardens and Arboreta (AABGA). Washington, D.C.: AABGA.

———. 1995. Lions and tigers and bears, oh no! In *Ethics on the ark,* ed. Bryan G. Norton, Michael Hutchins, Elizabeth F. Stevens, and Terry L. Maple. Washington, D.C.: Smithsonian Institution.

———. 1996a. Collaboration and conservation. *Public Garden* 11, 3 (Oct.).

———. 1996b. The design and use of moats and barriers. In *Wild animals in captivity,* ed. Devra G. Kleiman, Mary E. Allen, Katerina V. Thompson, and Susan Lumpkin. Chicago: University of Chicago Press.

Hardy, Donna Fitzroy. 1996. Current research activities in zoos. In *Wild animals in captivity,* ed. Devra G. Kleiman, Mary E. Allen, Katerina V. Thompson, and Susan Lumpkin. Chicago: University of Chicago Press.

Harris, Louis, and Associates. 1994. *Science and nature survey.* New York: American Museum of Natural History.

Hartley, L. P. 1958. *The go-between.* London: Penguin.

Hässlin, J. J., and Gunther Nogge. 1985. *Der Kölner Zoo.* Cologne: Greven Verlag.

Haworth, Philip, and Kathy Travers. 1993. Changing stripes. *ASPCA Animal Watch* (summer).

Hediger, Heini. 1950. *Wild animals in captivity: An outline of the biology of zoological gardens.* London: Butterworth.

———. 1955. *Studies of the psychology and behavior of animals in zoos and circuses.* London: Butterworth.

Heilborn, Adolf. 1929. *Zoo Berlin, 1841–1929.* Berlin: Raue.

Hertsgaard, Mark. 1997. Our real China problem. *The Atlantic Monthly* 280, 5 (May).

Hill, Arthur W. 1915. The history and functions of botanical gardens. In *Annals of the Missouri Botanical Gardens.* Vol. 2. St. Louis, Mo.: Botanical Gardens.

Holdridge, L. R., and J. A. Tosi Jr. 1972. *The world life zone classification system and forestry research.* Costa Rica: Tropical Science Center.

Horowitz, Helen Lefkowitz. 1996. The national zoological park. In *New worlds, new animals,* ed. R. J. Hoage and William A. Deiss. Baltimore: Johns Hopkins University Press.

Hoskins, W. G. 1955. *The making of the English landscape.* London: Hodder and Stoughton.

Hudson, Kenneth, and Ann Nichols. 1991. *The Cambridge guide to the museums of Europe.* Cambridge: Cambridge University Press.

Hunt, John Dixon, and Peter Willis, eds. 1988. *The genius of the place: The English garden, 1620–1820.* Cambridge, Mass.: MIT Press.

Hutchins, Michael, David Hancocks, and Theresa Calip. 1979. Behavioral engineering in the zoo. *International Zoo News* 25, 7 and 8 (Oct. and Nov.).

———. 1980. Behavioral engineering in the zoo. *International Zoo News* 26, 1 (Jan.).

Hutchins, Michael, David Hancocks, and Carolyn Crockett. 1978. Naturalistic solutions to the behavioral problems of captive animals. *Zoologische Garten* 54 (June).

International Union of Directors of Zoological Gardens/Captive Breeding Strategies Group. 1993. *The world zoo conservation strategy.* Brookfield: Chicago Zoological Society.

Janick, Jules. 1992. Horticulture and human culture. In *The role of horticulture in human well-being and social development,* ed. Diane Relf. Portland, Ore.: Timber.

Jennison, George. 1937. *Animals for show and pleasure in ancient Rome.* Manchester: Manchester University Press.

Jones, Grant, Jon Charles Coe, and Dennis R. Paulson. 1976. *Woodland Park Zoo: Long-range plan, development guidelines, and exhibit scenarios.* Seattle: Department of Parks and Recreation.

Jones, Steve. 1996. *In the blood: God, games, and destiny.* London: HarperCollins.

Jordan, Bill, and Stephen Ormrod. 1978. *The last great wild beast show.* London: Constable.

Kaplan, Rachel, and Stephen Kaplan. 1989. *The experience of nature.* New York: Cambridge University Press.

Kawata, Ken. 1991. Hediger who? A plea for historical perspective. *International Zoo News* 38, 4 (June).

Keeling, C. H. 1984. *Where the lion trod.* Guildford, Eng.: Clam.

———. 1985. *Where the crane danced.* Guildford, Eng.: Clam.

Kellert, Stephen R. 1981. *Knowledge, affection and basic attitudes toward animals in American society.* Washington, D.C.: U.S. Government Printing Office.

———. 1993. The biological basis for human values of nature. In *The biophilia hypothesis,* ed. Stephen R. Kellert and Edward O. Wilson. Washington, D.C.: Island Press.

Kellert, Stephen R., and Julie Dunlap. 1989. *Informal learning at the zoo.* Report to the Zoological Society of Philadelphia. ZSP.

Kirchner, J. W., and A. Weil. 2000. Delayed biological recovery from extinctions throughout the fossil record. *Nature* 404, 177.

Kisling, Vernon N. Jr. 1996. The origin and development of American zoological parks to 1899. In *New worlds, new animals,* ed. R. J. Hoage and William A. Deiss. Baltimore: Johns Hopkins University Press.

Kleiman, Devra G. 1992. Behavioral research in zoos. *Zoo Biology* 11.

Klös, Hans Frädrich, and Ursula Klös. 1994. *Die Arche Noah an der Spree.* Berlin: FAB Verlag.

Klös, Heinz-Georg. 1969. *Von der Menagerie zum Tierparadies.* Berlin: Haude & Spenersche Verlag.

Kohlmaier, Georg, and Barna von Sartory. 1986. *Houses of glass.* Cambridge, Mass.: MIT Press.

Kondonin Group. 1997. *National Australian school survey.* Perth.

Krutch, Joseph Wood. 1957. *The great chain of life.* London: Eyre & Spottiswoode.

———. 1961. *The world of animals.* New York: Simon and Schuster.

Leopold, Aldo. 1966. *A Sand County almanac.* New York: Oxford University Press.

Lewis, C. S. 1962. *The abolition of man.* New York: Collier.

Lewis, George "Slim," and Byron Fish. 1978. *I loved rogues.* Seattle: Superior.

Lindburg, Donald G. 1992. Are wildlife reintroductions worth the cost? *Zoo Biology* 11.

Link, Theodore. 1873. Zoological gardens. *The American Naturalist* 17.

Little, Charles E. 1995. *The dying of the trees.* New York: Viking.

Livingston, Bernard. 1974. *Zoo animals, people, places.* New York: Arbor.

Loisel, Gustav. 1912. *Histoire des ménageries de l'antiquité a nos jours.* Paris: Octave Doin et Fils.

Lovejoy, Arthur O. 1936. *The great chain of being.* Cambridge, Mass.: Harvard University Press.

MacKenzie, Deborah. 2000. Unnatural selection. *New Scientist* 2235 (22 Apr.).

Malmberg, Melody. 1998. *The making of Disney's Animal Kingdom theme park.* New York: Hyperion.

Mann, William M. 1949. *Wild animals in and out of the zoo.* Washington, D.C.: Smithsonian Institution.

Maple, Terry L., and Erika F. Archibald. 1993. *Zoo man: Inside the zoo revolution.* Atlanta: Longstreet Press.

Margulis, Lynn, and Dorion Sagan. 1986. *Microcosmos.* New York: Simon and Schuster.

Marr, Anthony. 2000. The tiger and the bear. http://www.HOPE-GEO.org.

Martin, Angus. 1995. Zoologist in the zoo. In *Zoo News* (Melbourne, Australia) 15, 4.

Martin, Paul S. 1984. Prehistoric overkill. In *Quaternary extinctions,* ed. Paul S. Martin and Richard G. Klein. Tucson: University of Arizona Press.

Massingham, H. J. 1935. A look at the zoo. In *Wonderful London.* London: Amalgamated Press.

Masters, Brian. 1996. *The evil that men do.* London: Doubleday.

McDonough, William, and Michael Braungart. 1998. The next industrial revolution. *The Atlantic Monthly* 282, 4 (Apr.).

McMaster, Lee. 1988. *An American perspective of landscape design in the development of American zoos.* Master's thesis. Cornell University.

Mitman, Gregg. 1996. When nature is the zoo. In *Science in the field,* ed. Henrika Kuklick and Robert E. Kohler. Osiris. Second series, vol. 11.

Morris, Desmond. 1964. *The response of animals to a restricted environment.* Zoological Society of London Symposia. No. 13.

———. 1965. *The mammals.* New York: Harper and Row.

Muggeridge, Malcolm. 1988. *Conversion: A spiritual journey.* London: Collier.

Mullen, Bob, and Garry Marvin. 1987. *Zoo culture.* London: Weidenfeld and Nicolson.

Nabhan, Gary Paul. 1995. The danger of reductionism in biodiversity conservation. *Journal of Conservation Biology* 9, 3 (June).

———. 1997. *Cultures of habitat.* Washington, D.C.: Counterpoint.

Nabhan, Gary Paul, and Sara St. Antoine. 1993. The loss of floral and faunal story. In *The biophilia hypothesis,* ed. Stephen R. Kellert and Edward O. Wilson. Washington, D.C.: Island Press.

Nabhan, Gary Paul, and Stephen Trimble. 1994. *The geography of childhood.* Boston: Beacon Press.

Nicholson, Thomas D. 1991. Preserving the earth's biological diversity. *Curator* 34, 2.

Nyhuis, Allen M. 1994. *The zoo book.* Albany: Carousel.

Olin, Laurie. 1996. Landscape design and nature. In *Ecological design and planning,* ed. George F. Thompson and Frederick R. Steiner. New York: Wiley and Sons.

———. 2000. *Across the open field: Essays drawn from English landscapes.* Philadelphia: University of Pennsylvania Press.

Orr, David W. 1997. *Earth in mind.* Washington, D.C.: Island Press.

Osborn, Fairfield. 1941. The opening of the African plains. *Bulletin of the New York Zoological Society* 44.

Osborne, Michael A. 1996. Zoos in the family. In *New worlds, new animals,* ed. R. J. Hoage and William A. Deiss. Baltimore: Johns Hopkins University Press.

Page, Jake. 1990. *Zoo: The modern ark.* New York: Facts on File.

Parsons, Christopher. 1996. The electronic zoo and other components of Wildscreen World. Conference Proceedings, Trends in Leisure and Entertainment. Maastrich, The Netherlands.

Powell, Anne Elizabeth. 1997a. Gardens of Eden. *Landscape Architecture* 87, 4 (Apr.).

———. 1997b. Breaking the mold. *Landscape Architecture* 87, 10 (Oct.).

Pratt, Nancy C. 1995. Letter to author. 28 Aug.

Preston, Douglas J. 1983. Jumbo, king of elephants. *Natural History* 92, 3.

Quammen, David. 1998. Planet of weeds. *Harper's Magazine* 297, 1781 (Oct.).

Raby, Peter. 1996. *Bright paradise.* Princeton: Princeton University Press.

Ralls, Katherine, Kristin Brugger, and Jonathon Ballou. 1979. Inbreeding and juvenile mortality in small populations of ungulates. *Science* 206

Reichenbach, Herman. 1996. A tale of two zoos. In *New worlds, new animals,* ed. R. J. Hoage and William A. Deiss. Baltimore: Johns Hopkins University Press.

Reilly, Charles. 1934. Letter to the editor. *Architect's Journal* (4 June).

Remnick, David. 1997. Future perfect. *The New Yorker* (20–27 Oct.).

Rensen-Oosting, Aleid. 1991. Noorder Dierenpark, Emmen. *International Zoo News* 38, 5 (Aug.).

Ripley, Dillon. 1969. *The sacred grove.* Washington, D.C.: Smithsonian Institution.

Robinson, John G. 1999. Losing the fat of the land. Editorial. *Oryx* 33, 1 (Jan.).

Robinson, Michael H. 1988. Bioscience education through bioparks. *BioScience* 38, 9 (Sept.).

———. 1993. Biodiversity, bioparks, and saving ecosystems. *Endangered Species UPDATE* 10, 3 and 4.

Sagan, Carl. 1985. Can we know the universe? Reflections on a grain of salt. In *Guest essays in science,* ed. Martin Gardner. Oxford: Oxford University Press.

Schama, Simon. 1995. *Landscape and memory.* New York: Knopf.

Segal, Jeanne, ed. 1988. *An animal garden in Fairmount Park.* Zoological Society of Philadelphia.

Seton-Thompson, Ernest. 1900. The national zoo at Washington. *The Century Magazine* 60, 1 (May).

Spearman, Diana. 1966. *The animal anthology.* London: John Baker.

Stearn, William T. 1961. Botanical gardens and botanical literature in the eighteenth century. In *Catalogue of botanical books in the collection of Rachel McMasters Miller Hunt.* Vol. 2. Part I. Pittsburgh: Hunt Botanical Library.

Stevens, William K. 1993. Zoos find a new role in conserving species. *New York Times,* 21 September.

Stott, R. Jeffrey. 1981. *The American idea of a zoological park.* Ph.D. thesis. University of California—Santa Barbara.

Streatfield, David. 1995. Regionalism in landscape design. In *Process Architecture* no. 126. Tokyo.

Strouhal, Eugen. 1992. *Life in ancient Egypt.* Cambridge: Cambridge University Press.

Sullivan, Robert. 1992. Trouble in paradigms. *Museum News* 71, 1 (Jan./Feb.).

Takacs, David. 1996. *The idea of biodiversity.* Baltimore: Johns Hopkins University Press.

Thomas, Keith. 1983. *Man and the natural world: Changing attitudes in England, 1500–1800.* London: Allen Lane.

Thomas, Lewis. 1979. *The Medusa and the snail.* New York: Viking.

Thoreau, Henry David. 1987. *The Maine woods.* New York: Harper and Row.

Toynbee, J. M. C. 1973. *Animals in Roman life and art.* London: Thames and Hudson.

Tudge, Colin. 1991. *Last animals at the zoo.* London: Hutchinson Radius.

———. 1995. *The day before yesterday.* London: Random House.

Turner, Jack. 1996. *The abstract wild.* Tucson: University of Arizona Press.

Tyldesley, Joyce. 1996. *Hatchepsut.* London: Viking.

Ulrich, Roger S. 1991. Effects of interior design on wellness. *Journal of Health Care Interior Design* (spring).

———. 1993. Biophilia, biophobia, and natural landscapes. In *The biophilia hypothesis,* ed. Stephen Kellert and Edward O. Wilson. Washington, D.C.: Island Press.

Van Gelder, Richard G. 1991. A big pain. *Natural History* 100, 3 (Mar.).

Vevers, Gwynne. 1976. *London's zoo.* London: The Bodley Head.

Wallace, Alfred Russel. 1869. *The Malay archipelago.* London: Macmillan.

———. 1876. *The geographical distribution of animals, with a study of the relationships of living and extinct faunas as elucidating past changes of the earth's surface.* New York: Harper and Bros.

Ward, Peter D. 1997. *The call of distant mammoths.* London: Copernicus.

Wedderburn, John. 1998. The zoos of China: A personal review. http://www.aapn.org/zoopage.html.

Weiner, Douglas R. 1988. *Models of nature: Ecology, conservation, and cultural revolution in Soviet Russia.* Bloomington: Indiana University Press.

Weise, R. J., and M. Hutchins. 1996. Evolution of the species survival plan. *Animal Keeper's Forum* 23, 2 (Feb.).

Werler, John E. 1975. Gorilla habitat display at Houston Zoo. In *International zoo yearbook.* Vol. 15. London: Zoological Society of London.

Weschler, Lawrence. 1995. *Mr. Wilson's cabinet of wonder.* New York: Pantheon.

Wharton, Dan. 1995. Zoo breeding efforts: An ark of survival? *Forum of Applied Research and Public Policy* 10, 1.

White, T. H. *The book of beasts: being a translation from a Latin bestiary of the twelfth century.* London: Cape.

Whitehead, Alfred N. 1967. *Science and the modern world.* New York: Free Press.

Wilson, Edward O. 1984. *Biophilia: The human bond with other species.* Cambridge, Mass.: Harvard University Press.

———. 1987. The little things that run the world (the importance and conservation of invertebrates). *Journal of Conservation Biology* 1, 4 (Aug.).

———. 1989. Threats to biodiversity. *Scientific American* 261, 3 (Mar.).

———. 1992. *The diversity of life.* Cambridge, Mass.: Harvard University Press.

World Resources Institute (WRI), International Union for Conservation of Nature (IUCN), and United Nations Environment Programme (UNEP). 1992. *Global biodiversity strategy.* Washington, D.C.: WRI, IUCN, and UNEP.

Worms, Mark K. 1989. Zoo exhibits and the role of zoo horticulture. In *International zoo yearbook.* Vol. 29. London: Zoological Society of London.

Wyburn, Theresa. 1966. In pursuit of useful knowledge. *Victorian Historical Journal* 67, 1.

Zuckerman, Solly, ed. 1976. *The Zoological Society of London, 1826–1976 and beyond.* London: Academic.

———, ed. 1980. *Great zoos of the world.* London: Weidenfeld and Nicholson.

INDEX